SO HELP ME, GOD

A U.S MARINE'S MEMOIR OF ABUSE, CORRUPTION, AND SELF-DISCOVERY

CHARLES VERLICE

Cover Illustration © 2025 Eagle and Globe Publishing LLC
Cover & Interior Design: Eagle and Globe Publishing LLC
www.eagleandglobepublishing.com

Published by Valor Books an imprint of Eagle and Globe
Publishing LLC
425 W Colonial Dr STE 303 #536 Orlando, Florida 32804
www.eagleandglobepublishing.com

ISBN: 979-8-9998379-0-5

Library of Congress Control Number: 2025945224

Printed in the United States of America Eagle and Globe
Publishing LLC: 2026

The Marines I have seen around the world have the cleanest bodies, the filthiest minds, the highest morale, and the lowest morals of any group of animals I have ever seen. Thank God for the United States Marine Corps
- Eleanor Roosevelt

For all the voiceless...

To the readers...

When I joined the United States Marine Corps back in 2011, I had the same vision and dreams every eager young man has: to serve the country and become a hero. Of course that was nothing more than phony tough guy bravado and reality set in pretty fast for me during my service in the USMC. People reading this may form their own opinions and that is fine. I spent many years holding all this in thinking that I would take it to the grave with me and nobody would ever have known anything about my story. This story, a cautionary tale for you young men (and young women), is not meant to show any disdain for the USMC. I am proud, even to this day, that I am a Marine and I will be one until the day that I draw my last breath. I just felt deep down that my story was unique and not like your typical Marine Corps novel of guns, explosions, beer, and death. This story is one that is very personal to me and I am willing to bear it all on these pages so that perhaps someone, somewhere may find inspiration or guidance in my words. If nothing else, I hope this book helps you to realize the sacrifice that young men and women who join the service deal with everyday. The voiceless... the silent. Thank you all for taking the time to read my story and of course...SEMPER FI, Marines!

-Goodnight Chesty Puller, wherever you are...if you know, you know!

CHAPTER 1
A CIVILIAN DIES, A RECRUIT IS BORN

It was pitch black, O'dark thirty as they say in the military, as the bus I was being transported on pulled up to the gates of Parris Island. We had just stopped at a McDonalds before reaching the gate where everyone had one last cigarette, a burger– if nerves permitted, and a quick prayer. It would be our last contact with the outside world for the next three months. As the guards at the gate checked all our ID's, one barked out, "Who is Verlice?"

Why is he calling out my name?

There were twenty other people on the bus with me, so I thought it was weird that I was the only one being singled out.I hesitated for a moment and returned, in a soft and timid voice.

"Here."

"Yea, you're not going to pass the drug test!", he snarled in anger, throwing the IDs on the floor of the bus before departing back to his post.

There I was, not even through the gate and I was already being targeted, I assumed, for the fact that I had long hair, and looked as though I partook in recreational drug use. That really set the tone for my stay on the exclusive Parris Island and made me even more nervous about the journey I was about to embark on. The bus pulled slowly through the gate and continued to the next destination – the infamous yellow footprints.

The bus drove for what seemed like an hour along a dark and winding road as it slowly snaked its way to the receving buidling. I remember having an overwhelming feeling of nervousness beginning to creep up; it was like the type of feeling one gets before they go up to give a big speech in front of their entire class. I looked outside of the window and could only see dark swamp... there were no buildings! I remember thinking to myself, *Jesus, how long is this road?*

Eventually we rounded into the main part of the base but because it was so dark outside, I couldn't really see much of anything other than just roads and unmarked government buildings. One thing I did see, illuminated by the soft orange glow of streetlights, was these metal pipe groupings which seemed to be suspended in the air by poles. All the pipes were above ground and just zigzagged through the Island. Steam could be seen billowing from a few of them, and I remember thinking that it was somewhat unusual as one does not normally see pipes above ground. It was just one of the many things that immediately made Parris Island a ghostly and bizarre place to be.

It was not long before the bus driver, a decent enough fellow, told us all to put our heads down between our knees. They did that, I'm assuming, to disorient us more as to our surroundings and where we were and what was about to happen.

Let me tell you...it worked! Not only did it work on the mind, but it also worked on the body. I began to feel that nervous nausea creep up on me and sort of began having second thoughts about my decision to join the United States Marine Corps. As the bus came to a stop, my heart started to pound even harder and even though nobody could see where we were, we all knew.

Parris Island is one of two of the Marine Corps Recruit Depots (MCRD) where any potential recruit residing east of the Mississippi River crazy enough to sign a USMC contract is sent to for boot camp (the other being in San Diego). Every recruit starts their journey as a potential United States Marine at the very same spot as thousands of other Marines – at the yellow footprints. It's not only the first time a recruit will stand at the position of attention, but it is the beginning of a rite of passage! I remember getting goosebumps all down my arms as I sat on the bus with my head between my legs. I could practically feel the spirit of all the Marines – past and present – who have fought and died for the causes of the United States since November 10, 1775.

We had indeed arrived at the yellow footprints, and I knew it would not be long before an angry drill instructor would break into the bus and scream for us all to run off as fast as we could. I had no idea what to expect or what was going to happen. Thoughts of torment, abuse, and even death began to run through my mind since after all I was going to be an infantryman if I survived bootcamp. *Would I ever be normal again? Would I ever get to make it home again?* All I knew was that I had to at least make it through the next thirteen weeks and then I could figure the rest out later.

After what seemed like hours, I heard the bus door open with a deafening creak. "GET OFF MY DAMN BUS RIGHT NOW!", the faceless voice bellowed with great force. Everyone on the bus screamed simultaneously, "YES SIR!!" I shot up from my seat to see everyone filing off as fast as possible. I quickly got off the bus and ran straight into the brisk December evening air where everyone was running as fast as they could to the yellow footprints. The Drill Instructor – some nameless Marine in a smokey the bear hat campaign cover – was screaming "15, 14, 12, 11, 10, 9, 8, 7, 6, 5, 4, 3, 2, 1…!" It was our countdown to get onto the yellow footprints on *their* time…which was not much time at all.

By the time he screamed "1!", I was on my spot on the yellow footprints, standing at the position of attention with nothing but the clothes on my back – an old red button up which I considered my "lucky shirt", a pair of jeans, a pair of old new balance sneakers, and my wallet which contained roughly a dollar and some change in cash. It was all I had in the world – all I had left – and all I left behind as the last seconds of my civilian life came to an end.

The drill instructor, who was simply just what they called a receiving drill instructor and not one of the permanent drill instructors who would train us, was pacing back and forth as we all stood there shivering at the position of attention on the yellow footprints outside of the receiving barracks. I can still see it now; I can still see exactly where I was standing. I could already hear tons of screaming from other recruits on the island who had arrived earlier. It was already chaos! The distant "AYE SIR!" screams which floated through the island like a ghostly echo sent shivers down my spine.

The drill instructor started his famous speech in that distinctive gruff and rough bullfrog voice every USMC drill instructor has. "YOU ARE NOW ABOARD MARINE CORPS RECRUIT DEPOT PARRIS ISLAND, SOUTH CAROLINA AND YOU HAVE JUST TAKEN THE FIRST STEP TOWARDS BECOMING A MEMBER OF THE WORLDS FINEST FIGHTING FORCE – THE UNITED STATES MARINE CORPS."

Jesus, I was really here and this was really happening

I stood there at the position of attention trying not to shiver as the cold swampy winds whipped across my face. Nothing went through my mind other than the angry drill instructor who continued to pace back and forth while we all stood at the position of attention.

It seemed almost like a dream as the drill instructor continued his speech which included going over the UCMJ (Uniformed Code of Military Justice) and other customary details. My head was locked straight forward, fighting off the urge to look around at my surroundings. I did not dare do so – I don't think anyone did or ever would.

The Drill Instructor (DI) continued, "THE MARINE CORPS' SUCCESS DEPENDS ON TEAMWORK. TEAMWORK, THEREFORE, IS AN ESSENTIAL PART OF YOUR TRAINING HERE AT PARRIS ISLAND. STARTING

NOW, YOU WILL TRAIN AS A TEAM. YOU WILL LIVE, EAT, SLEEP, AND WORK AS A TEAM. THE WORD, "I", WILL NO LONGER BE A PART OF YOUR VOCABULARY. YOU WILL SAY THIS RECRUIT, THAT RECRUIT, THESE RECRUITS…DO YOU UNDERSTAND!?"

We all screamed, "YES SIR!"

"YEA RIGHT! DO YOU UNDERSTAND?!!"

"YES SIR!"

As it seemed to get colder with the chilly nighttime air picking up, the receiving DI finished his speech, "TENS OF THOUSANDS OF MARINES BEGAN OUTSTANDING SERVICE TO OUR COUNTRY ON THE VERY FOOTPRINTS WHERE YOU ARE STANDING. ARE YOU READY TO CARRY FORWARD THEIR TRADITION?"

"YES SIR!"

"YEA RIGHT, ARE YOU READY?!", he screamed again wanting more volume from the lot of us.

"YES SIR!!!", we screamed louder that time, sending echoes throughout the island.

"FOLLOW ME UP TO THOSE HATCHES!"

"YES SIR!"

We then ran up to the receiving building front hatches (doors) where above them read in bold: THROUGH THESE PORTALS PASS PROSPECTS FOR AMERICA'S FINEST FIGHTING FORCE -UNITED STATES MARINES.

THE FOG OF WAR

The fog of war is a term used by the Marine Corps (and the Army) to describe the sort of fatigue, confusion, sickness, pain, and misery experienced by Marines and Soldiers in a combat situation. You're really running blind, fighting complacency, and in a state of total confusion during the "Fog of War". Well, the first night on Parris Island certainly qualifies as a Fog of War-type situation simply because we were thrust into a world of bizarre and craziness that we, until entering those front hatches of the receiving building, were completely ignorant to.

Unlike other branches including the Air Force and maybe even the Army, Marine recruits are not given the luxury of a night of sleep upon their arrival. There was no beauty sleep that first night, my friend; they kept us up for the first two days running here and running there while they introduced us to the new world of the USMC.

I still remember, vividly, the first thing that I noticed upon entering the receiving building was… the smell. The smell was God awful! It was like the most pungent and potent government-issued cleaning solution – bleach and pine sol – smell I had ever had the displeasure of smelling. It permeated my nose and immediately gave me a migraine. I have smelled a lot of bad smells in my day, but the smell in the receiving building barracks on Parris Island was one of the worst and will never leave my memory.

The first few moments of this experience consisted of writing our names on a little piece of paper/laminated type of thing (sort of like a temporary name tag since the uniforms didn't have our names on them for the first few weeks). They then had us running all around like chickens with our heads cut off, and of course, there were lots and lots of screaming and yelling. The first thing which became quite apparent was that no matter how loud we screamed, "AYE SIR!!", it was never loud enough, and it was not uncommon to have to scream it back upwards of twenty times as loud as possible before they would be satisfied.

With that being said, I do distinctly remember thinking to myself, *you know, this isn't really that bad so far, I might just be able to get through this without any problems. Sure, they're yelling and screaming, but honestly, it's not too bad.* That, however, was a drastic false assumption I was making as the receiving drill instructors were not as hard on us and what they were giving us was a watered-down version of what we would be getting for real at the end of the week.

One of the first things we did in the receiving building was make our one and only phone call home. Standing in front of a phone box on the wall – literally something from the Vietnam War era – I made my call home reading only from the prompt which was crudely taped onto the door of the prehistoric device. This detail may not seem that important, but it is. In this day of age where we are so used to technology and all the comforts of society, seeing an old, rusty phone decades old really played on my psyche and made me feel that much more cut off from the world and was just one of the many tactics the Marine Corps used to begin to break us down piece by piece.

I made haste in dialing the phone number which I had memorized for that one and only occasion. My mother picked up on a landline phone which was installed specifically for the purpose of receiving that one and only call and said nothing as I said aloud, with a drill instructor breathing down my neck, "Hello. I have arrived safely at Parris Island. Please do not send me anything bulky in the mail. I will write to you in a few days. Thank you for your support. Goodbye for now."

The whole thing lasted ten seconds; she didn't even have time to get a word out before I was pulled away from the phone and herded along with a large group of other screaming recruits into a small room where we would receive our first haircut – the bootcamp cut!

We were sardine-packed with at least 50-60 recruits inside that barber room, and I was beginning to feel very claustrophobic. We had to stand what they would call "nut-to-butt" with each other which meant my face was jammed into the back of another recruit's head and vice versa. It was hot and everyone was beginning to feel the pain by that point. Tired, aching, sweaty – yea it was beginning to dawn on me now that things were going to only go downhill from there. As we were standing there for what seemed like an hour, I remember seeing some kid, a kid who ended up being in my bootcamp platoon with me named Moline, looking very pale, sweaty, and wobbling back and forth. I'm not sure if he was about to spew chunks or pass out but he looked as though he was going to do both simultaneously.

We got buzzed and filed out into a classroom looking space where we gave up all our possessions. For me it was my wallet and the clothes on my back. We were then filed out to receive all our uniforms and gear – about 80 pounds shoved into two seabags – as well as other toiletries. Don't forget all the screaming! The screaming never stopped even for a second. We were running everywhere and walked nowhere and most of the time we had to repeat the drill instructor's orders multiple times for one reason or another.

We then ran up three flights of stairs, about 10 times because people were not moving fast enough for the drill instructor, and then were staged, with about 150 guys, in a large, empty squad bay. The smell was much worse up there and I was beginning to feel sick and tired as things got more hectic. They were ordering us to sit in certain spots with certain groups of recruits further making the situation confusing. Looking back in hindsight, they were sectioning us off into what platoons we were going to be, but because of the Fog of War, I really didn't know what they were doing or what I was supposed to be doing. It was honestly just a bunch of chaos.

I remember distinctly wanting to force myself to shit. My mother and father had both served in the United States Air Force and told me that they both had trouble shitting when they got to basic training. My mother, who had a knack for scaring the crap out of me with her stories, told me that one girl she was in basic with had to get an enema because she couldn't defecate. I did NOT want that happening, so I made it my mission to try and force a shit – even though I didn't have to go. I felt like if I could just squeeze even one pebble out that somehow, I would be good for a few days, and the dreaded enema would not be needed.

There was some downtime there while we were waiting in the empty squad bay and even though Drill instructors were yelling all over the place, recruits were screaming at the top of their lungs, and there was just overall confusion, some recruits were running into and out of the head (bathroom) for one reason or another. I requested permission and ran to the head to try and force a shit. On the run to the head, I glanced around the open squad bay and estimated that there were at least three groups of sixty to seventy recruits sitting Indian-style on the deck.

The head looked as though it had not been updated since 1942 and while that may be an exaggeration, I'm sure it was not too far from being accurate. It was drab, cold, and surrounded by cinderblock walls with only two doorless stalls. I sat on that cold porcelain thrown and prayed to God to let me drop a turd, but I was unfortunately unable to. The screams from all the drill instructors and recruits echoed through the squad bay and into that weird bathroom like something out of a horror story, and as a result, my asshole was wired tight. I really cannot put it into words what it was like sitting there fighting for a turd to come out while listening to those deafening screaming echoes. It was just creepy.

Running back to where the rest of my group of people were sitting, I passed by several large puddles of puke from various recruits that had been unable to make it to the head. Shortly after I got back to my group, the DIs had made several recruits pass out box chows - known as "bag of nasties" - to all the recruits who were just sitting around the squad bay. I had a hard time eating because my stomach was not feeling great, and my head was pounding.

I remember feeling as though I was going to vomit as I slowly dug into the "bag of nasty" where I pulled out an old sub sandwich with the driest bread which hugged a slice of cheap ham and cheese. In addition, there was a bag of the most generic chips, a raisin cookie or box of raisins, a pack of Gatorade powder which we were not allowed under any circumstance to put into our canteen, and – get this – a hard-boiled egg. I don't know how I didn't lose it right there, but somehow, I was able to force down some of the chips and dry Gatorade powder which I made into small sips of flavored water by swishing swigs from my canteen in my mouth.

At some point shortly after forcing that slop down my throat, I remember a random Drill Instructor entered onto the receiving barracks floor and started screaming for us all to look at him. "ALL YOU BITCHES LOOK AT ME RIGHT NOW!", he screamed, the veins poking out of his neck like over taught bungee cords. "I'M GOING TO GIVE YOU A PREVIEW OF WHAT YOUR REAL DRILL INSTRUCTORS ARE GOING TO DO TO YOU IN A WEEK!" He was screaming crazily, causing everyone to tense up and lock their eyes in his direction. "YOU SCUMBAGS! I WILL MAKE YOU PUKE, SHIT, PISS, AND PASS OUT ALL WITH ONE COMMAND! YOU WILL BE WISHING YOU NEVER SIGNED THAT CONTRACT, FUCKERS!" He went on taunting us for some time before eventually departing down the stairwell. It was clear that they meant absolute business on Parris Island.

As the sun began to come up, all the recruits, including myself, were herded around over the receiving barracks from room to room for various reasons. Finally, we ended up in a classroom type space with maybe 20 or so desks. They told us to put our heads down, which was a welcome order because I could close my eyes if for only a few minutes. I remember trying so hard to force my eyes open as I was fighting off the urge to sleep. That moment in the classroom was what they called the "moment of truth", and it was where they called everyone up one at a time to go in front of an officer – some captain – and spill their guts on anything they may have lied about at MEPS (military entrance processing station) or basically proclaim to them that they wanted to quit. It ran through my mind, as I'm sure it did most, to tell the officer, "Hey, look man... I want out of this crazy shit! I want to go home!" Of course I didn't and simply I told the officer that I had not lied about anything and was ready for the rigors of bootcamp...*so help me, God!*

I remember the rest of that day being mixed with lots of running around and being issued a few more things including the rifle. We marched poorly, looked like shit with our uniforms worn incorrectly and our covers (hats) pulled down over our head way too much, and lacked the overall discipline required to be a United States Marine. We all wore sneakers instead of boots (in the Marines we call sneakers go-fasters) and a glow belt (reflective belt) strapped diagonally across our bodies. That was done to let everyone know that we were brand new and didn't know shit yet.

At some point during the first full day, one of the receiving drill instructors started throwing out ponchos in the middle of that empty squad bay we had spent much of the previous night in. It was very unorganized, and recruits were basically running haphazardly to grab one and then get back in formation. By the time I made my way to where the DI had been throwing out the ponchos, they had all been taken. I began to internally freak out a bit because I was the odd one out and I had no intention of being singled out due to some missing gear, so I decided to muster the courage to approach and ask the receiving drill instructor for a poncho. He said he would get me one later. As the day went on, however, he never gave me a poncho, and I was too afraid to ask again him, quite frankly. Even the receiving drill instructors were not approachable and given the fact that I had already been hit in the back of the head by one for moving "too slow" earlier that morning, I really did not think it was wise for me to ask them for anything else. That ended up being a huge mistake on my part.

At some point early the next day we were instructed to run about half a mile with all our gear and seabags slung over our backs to the barracks where we would call home for the next three months. I remember how hard that run was because of the amount of weight I was carrying and the speed at which they wanted us to move.

Our barracks were on the third deck (floor), and we all had to run up those flights of stairs several times because people were moving too slowly and not making it on line by the end of the countdown. Getting on-line meant that we were all to be standing at the position of attention in front of the bunks (racks) - no moving, talking, or looking around.

The instructor would count us down from 60 seconds and when they would reach zero we were all expected to be on line. If every single recruit was not on their spot on line in front of a rack standing completely motionless at the position of attention, then it was right back outside running down the stairwell to do it all over again.We did that many times over and over.

Finally, we all made it up on the third deck and on line by the end of the countdown and the DIs instructed us to clean the squad bay spotless. They taught us how they wanted the barracks cleaned, how things were supposed to look; they taught us how to make a tight rack (make the bed to their standards), and we marched to the chow hall for the first time in two days since arriving. The receiving DIs were deceptively calm during and only yelled when they were speaking to us or if a recruit seriously warranted an ass chewing.

That night I fell asleep on my rack and slept the whole night without having a single dream – just pure exhaustion. The next morning, I remember we had our first drop out; the kid didn't even make last a week on Parris Island. I remember he was some black kid – don't know his name – who suddenly "couldn't find his uniform". The receiving DIs had us all on our hands and knees searching for the recruit's uniform while the kid just stood on line looking as if he was going to burst into tears. Clearly it would have been impossible for him to misplace his uniform in the middle of the night after taking it off and hanging it on the rack before lights out and it was obvious that he just couldn't hack it. He was gone that same day.

The rest of the week was sort of uneventful and consisted of being shuffled around here and there – like medical and dental – to get all our checks and everything else done before training could officially start at the end of the week. I remember getting the god-awful penicillin shot in my ass cheek. We called it the peanut butter shot because it felt like a thick glob of peanut butter wedged inside of the ass cheek. It hurt so badly for several days to the point where it hurt to even walk.

During the first week, the receiving drill instructor taught us how to march from here to there – like very rudimentary marching. They taught us the basic fundamentals of the M16-A4 service rifle; and they taught us several other basic things such as how to make a rack, how to stand in formation, and how to clean the head. Basically, they were just skimming the surfaces so that when we met our real drill instructors on what they call "Black Friday", they could get straight to the "fucking us up" without having to bother with the basics.

Another thing that the receiving drill instructors did was tell all types of alarming stories about what our real drill instructors were going to do to us. I remember the guy saying that if you do this or do that, you'll probably catch a "book" or elbow to the face. I wondered if he was being honest or if it was all just bullshit.

Receiving week, or the very first week of Bootcamp, which is not part of the "training schedule", ended with the IST (initial strength test) which consisted of the pull ups, sit ups and just a 1.5-mile run (half of the 3 miles required to graduate bootcamp). I was shocked at how poor my runtime was compared to how it was right before I left for bootcamp. I ran a 9.5-minute mile and a half before leaving and a 12.30 (barely making the score) at the end of receiving week. I attributed my poor run time to not only weakness but also the penicillin shot which was still throbbing in my ass cheek.

Immediately after the IST, we were marched back to the barracks, took a quick 30 second shower, and then were seated in two columns in the middle of the squad bay to await our drill instructors. Once we were all seated, the receiving drill instructors disappeared and were never seen again.

From there on out, it was going to be our real drill instructors who would be in charge of us. I remember sitting crisscross on the ground and seeing our company commander of Mike Company, Captain Midgeo, come out to give us our first speech before bringing out our real drill instructors. Many thoughts raced through my head and I had no idea what was about to befall me in just a matter of moments.

CHAPTER 2

NEVER QUIT AND NEVER GIVE UP!

2A – PHASE I

Before we get into the real fun stuff, let's go over a few basic things so that you as the reader will have a basic understanding of what I am talking about as some of this stuff will be very important later. First off, Marine Corps Bootcamp is broken up into sections over the course of thirteen weeks. A recruit has to go through receiving week and then onto Phase I, Phase II, Phase III, and then finally ending with Marine Week. Phase I lasts about three to four weeks and is basically the worst part of the whole thing. This is the period of time when the drill instructors are on their absolute worst behavior and try anything in their power to get a recruit to break, lose their bearing, or quit. They are trying to weed out the weak ones and the people who can't hack it.

For the most part, anyone who isn't going to become a Marine will generally not make it past Phase I and it was very rare (barring something medical) for someone to be dropped or quit if they made it past Phase I – although it did happen. Phase II is where a recruits starts to learn some more things such as shooting the rifle, the gas chamber, and several other things. Much of Phase II is spent on the rifle range in the rifle range barracks. Phase III is where things start to come together and recruits participate in final drill, BWT (basic warrior training), and finally the crucible before ending up at the Iwo Jima Monument to receive the EGA (eagle, globe, and anchor) emblem signifying that a recruit had made it through bootcamp and became a Marine. It seems like a long way from Phase I, and it is.

Next, let me inform you about the rank structures. So, to start, recruits have NO rank. I know the Army calls their recruits "private" – which is the lowest enlisted rank of both Marines and Soldiers – but the Marines don't do this. A recruit must become a Marine first before they even rate (or earn/ deserve) to be called a private. Everyone in Marine Corps bootcamp is a simply a "bitch-ass" recruit until the day they earn the title of United States Marine. If one does make it through USMC bootcamp the rank structure is as follows:

E-1: Private

E-2: Private First-Class

E-3: Lance Corporal

E-4: Corporal

E-5: Sergeant (NCO)

E-6: Staff-Sergeant (NCO)

E-7: Gunnery Sergeant (NCO)

E-8: First Sergeant or Master Sergeant

E-9: Sergeant Major or Master Gunnery Sergeant

*** Special** E-9: Sergeant Major of the Marine Corps (highest one can go as enlisted)*

The officer ranks are as follows:

O-1: Second Lieutenant

O-2: First Lieutenant

O-3: Captain

O-4: Major

O-5: Lieutenant Colonel

O-6: Colonel

O-7: Brigadier General

O-8: Major General

O-9: Lieutenant General

O-10: General (Four Star)

Some of these ranks will be quite important later, so remember to keep them in mind!

Okay, back to the shit storm...

So, there I was sitting crisscross in the middle of the squad bay with everyone else in the platoon. We all sat still and erect as possible as the captain went over some formalities about what to expect during the bootcamp training process. After several moments he announced, "I will now introduce you to *YOUR* drill instructors." All at once, three stone-faced drill instructors came marching out of the drill instructor hut (DIs room attached to the squad bay where there was always a drill instructor on duty 24-7).

The DIs marched out right in front of the recruits seated in the center of the squad bay, halted, made a right face movement, and stood at the position of attention. The captain then approached them, and they all raised their right hand and gave the drill instructor oath – which is essentially a small oath promising to train the recruits to the best of their ability to become basically trained Marines. Once that speech was completed, the senior drill instructor... *(break)...okay, time for a bit more housekeeping for reference.*

The drill instructor team is typically made up of three to five drill instructors which each have their own specific duties. They can all fill in for each other when needed, but this is pretty much how it's broken up:

Senior Drill Instructor (Senior DI): *the most senior of the drill instructors. He is the Marine who is in charge of the other drill instructors and also of making sure that the training schedule is on track. They also make sure the other drill instructors stay in line (to some degree) and don't accidentally kill a recruit or go way too far with their punishments. They also perform final drills and things like that. They are basically like dad. A recruit is supposed to be able to bring up any discrepancies to them to be handled accordingly.*

Heavy Hat: *the next senior drill instructor. They oversee all things drill, rifle movements, and other drill-type things. They also do a lot of the marching as well as provide discipline.*

Kill Hat *(may have more than one): the least senior of the drill instructors. These DIs one and only job is simple – to kill recruits. They are in charge of all the punishments and IT (intensive training). They are machines, animals, monsters, whatever you want to call them.*

In addition to that, there is also the **Chief DI** *who oversees several platoons and their drill instructors and are not seen as much as they perform more "administration tasks".*

It's also important to note that all drill instructors start off as Kill Hats so when needed, each drill instructor can revert back to their Kill Hat days and give a recruit a good ass chewing or beating.

Okay…got that? Back to the shit…

…faced the recruits, took a step forward and began his drill instructor speech. He spoke in a loud and commanding voice which got everyone's attention fast. "SIT UP STRAIGHT AND LOOK AT ME!" Every recruit stiffened up even more, fixing their gaze on this drill instructor. My ears seemed to open as he continued.

"I am Staff Sergeant Fierro, and I am your Senior Drill Instructor. I am assisted in my duties by: Drill Instructor Staff Sergeant McHolland and Drill Instructor Staff Sergeant Stubb…", each drill instructor stepped forward as their name was called.

"…OUR MISSION…", both Staff Seargent McHolland and Stubb snapped to parade rest in unison as Staff Sergeant Fierro continued, "…is to train each one of you to become a UNITED STATES MARINE. A Marine is characterized as one who possesses the highest military virtues. He obeys orders, respects his seniors, and strives constantly to be the best in everything that he does. Discipline and spirit are the hallmarks of a Marine. Each of you can become a Marine if you develop discipline and spirit. We will give every effort to train you, even after some of YOU have given up on yourselves! Starting now, you will treat me and all of the Marines with the highest level of respect, for we have earned our places as Marines, and we will accept nothing less than that from you. We will treat you as we do our fellow Marines: *with firmness, fairness, DIGNITY, AND COMPASSION!* Physical or verbal abuse by any Marine or recruit will **NOT** be tolerated. If anyone should abuse or mistreat you, I expect you to report such incidents immediately to me, or one of my drill instructors. My drill instructors and I will be with you every day, EVERYWHERE YOU GO! I have told you what my drill instructors and I will do. From you we DEMAND the following: you will give 100% of yourself, at all times. Obey all orders, quickly, willingly, and without question. Treat all Marines and recruits with courtesy and respect. You will not physically abuse or verbally threaten any Marine or recruit. Be completely honest in everything you do. A Marine never lies, or cheats. Respect the rights and property of all others. A Marine never steals. You must work hard to strengthen your BODY, your spirit, AND YOUR MIND! Be proud of yourself and the uniform you wear. Above all else, *NEVER QUIT, AND NEVER GIVE UP.* We offer you the challenge of Recruit Training – and the opportunity to earn the title, **UNITED STATES MARINE!**"

The whole thing seemed unreal to me at that moment. Only a few months prior I was sitting at a high school desk daydreaming about making out with the girl next to me and now I was getting ready to undergo the rigors of USMC bootcamp. I had no doubts in myself as I knew that nothing on earth or in Hell could ever make me quit, but I would have been lying if I said that I wasn't a bit scared about what could potentially happen in the coming weeks.

Staff Sergeant (SSGT) Fierro bellowed, "Now when I tell you

to, you will quickly rise to your feet, do you understand that?!"

"YES SIR!", the whole platoon screamed loudly.

"YEA RIGHT...SCREAM IT!"

"YES SIR!"

The squad bay rumbled as we screamed back louder.

"READY...MOVE!" We all screamed, "YES SIR", as we rose to our feet and began shaking our lower extremities to allow the blood to flow back into them.

He started to count again, "10, 9, 8, 7, 6, 5, 4, 3, 2, 1... SCREAM DONE SIR!"

We all screamed, "DONE SIR!"

Eventually we were instructed to get to our place back on line at the position of attention to await further instruction. The screaming rose to new levels that were not typical of the first week of bootcamp which indicated to me that shit was starting to hit the fan. The other two drill instructors were running around screaming at recruits to get on line and to scream as loud as they could and would scream at them to stop moving after the counts. Even moving a centimeter after the Drill Instructor reached 0 in the counts was grounds for not only punishment on yourself but on the whole platoon.

Once back on line, the DIs messed with us for several moments by making us scream certain things over and over before eventually instructing us to hold out the blue money valuable bag (a small bag which contained something called an Easy pay debit card) so they could make sure we all had it. The senior DI then instructed us to grab our canteens and hold them in our forearms parallel to the deck in a fashion which was supposed to resemble a rifle at right-shoulder arms. At that point a few recruits were already beginning to get singled out for not moving quickly enough, not screaming loud enough, or not having what they needed when it was requested. Every time the senior DI ordered us to bring something on line, we had to hold it out in front of us straight and parallel to the deck so that they could count us off.

Once they counted, we put it down back by our sides. If a recruit didn't have what they demanded in their hand when the countdown ended, then all hell would be released on them, and it was one of the quickest ways to get singled out. After a few moments, the senior DI instructed us to do a left face and march back to the quarter deck and sit down crisscross in two separate formations. These were the first five or ten minutes of meeting our new drill instructors and it was beginning to set in for me how serious, intense, and vicious these DIs could be. I'm sure everyone else had the same thoughts too.

Once we got onto the quarter deck, we sat down and SSGT Fierro instructed the other two DIs to go into the DI house so that he could address us without interruption. It was interesting to see how the more junior DIs instantly responded to the Senior DI by screaming "Aye Senior Drill Instructor!" as they quickly disappeared into the DI house.

Those "sit down" times (called school circles) on the quarter deck with the Senior DI were honestly some of the more laid-back moments even though he still spoke in a demanding and loud, but relatively civil manner. He would let us know what we were going to expect for that day or training event and would sometimes answer various questions recruits had. That was the first sit down he was having with us and I'm sure the conversation had something to do with what was going to happen the rest of the day and perhaps the week. I honestly don't remember what he said that day, but what I do remember was at some point during that lecture, the blinds covering the window to the DI house ripped open causing all the recruits to dart their glances towards it.

SSGT Stubb, who was by far one of the hardest and most intense DIs to ever walk the Island, was standing at the window screaming and pointing at us. Copious amounts of frothy spit flew from his mouth and collected on the window and slowly dribbled down as he presumably screamed profanities and threatened to kill us all. Everyone looked frozen in fear as we realized that these DIs meant business.

"DRILL INSTRUCTOR…STOP!!!", screamed SSGT Fierro. The blinds closed and we quickly snapped our attention back to SSGT Fierro. As he continued to demand of us what we would be doing in the coming hours and days, I felt in my guts that the next few weeks were going to be some of the most trying I had ever been through.

2B – Phase I weeks 1-4

My initial goal when starting bootcamp was to not be singled out at all by any Drill Instructors. Any platoon on Parris Island had between 60-80 recruits give or take, and even though the DIs were with us every day and everywhere we went, it was impossible for them to give "special treatment" to every recruit. I would estimate that roughly 60-70 percent of recruits don't get that "special treatment" and sort of blend in – like wallflowers; and while every recruit goes through the same rigors of bootcamp with the training schedule, not every recruit had those memorable and, what some may call, traumatic moments, with the DI's.

It was not uncommon for the DI's to not even know some of the recruits' names after 13 weeks just because they never got that one-on-one attention with them. Being unknown to a drill instructor was a good thing and something that a recruit wanted to strive for. Movies like *Full Metal Jacket* and many other Marine Corps-themed films are usually very inaccurate, having simply been dreamed up by some ignorant dumb ass with deep pockets. The movie *Full Metal Jacket* was like nothing I experienced on Parris Island except one part in the movie where Gunnery Sergeant Hartman says to Joker, "I've got your name, now I got your ass!" That, I found out the hard way, was true!

When a DI got a recruit's name, you can bet your ass that they would be their go-to whenever they wanted to fuck with someone, were having a bad day, or just simply wanted to take out some aggression on a helpless recruit. Everything a recruit did would be more scrutinized simply because they were on the DI's radar. Once they had a name, bootcamp would become a whole lot more difficult than it already was. Going into first phase, I tried to make it my goal for them not to get my name… and I failed!

The screaming I heard on that first day, once we got back online after SSGT Fierro finished talking to us, was like nothing I had ever heard before in my life. Certain recruits were being "voluntold" to do things such as being appointed the scribe (recruit who writes important things down), the big gear recruit (recruit who was in charge of putting all the issued shower gear into a closet we called Big Gear Closet", and several other pointless billets such as "squad leaders". They were beginning to strip people down bit by bit – big or small, it didn't matter. One guy (the Big Gear recruit) had to carry all the gear, which probably amounted to 70-80 pounds of shampoo, shaving cream, aftershave, tissues etc. on his back as he ran up and down the squad bay while getting screamed at.

His screams sounded like someone who had been hit by a car and lay dying on the side of the street, and was reminiscent of a girl screaming for her life. That really stands out to me even today because it's not the type of noise one hears on a daily basis. I just stood on line with my head straight forward wanting to look but dared not to lest I be called out by a DI and suffer the same fate. It honestly sounded like they were killing the poor guy. At that moment, I remember the kid in front of me, a guy called Backer, pissed himself. I saw he was shaking while a puddle of urine slowly began to form around the base of his feet. I remember thinking, *Goddamn, this kid just pissed himself. This is crazy!*

Backer was the skinniest kid in the platoon and was a major target for the DIs right out of the gate due to his overall appearance. He began to get singled out immediately by the DIs for moving too slow, not screaming loud enough, and overall not cutting the mustard. He earned the nickname "Ghandi" for his uncanny resemblance to Mahatma Ghandi. It was clear from day one that the kid was not going to make it through bootcamp; he just didn't have what it took.

The first week of phase I was pretty much just a constant haze fest. The DIs looked for any reason to "kill" a recruit and anyone was fair game. Every moment was hectic - from waking up in the morning, taking a piss or shit, eating, learning the marches, studying knowledge (classes on USMC knowledge), and everything in between. Every moment during first phase, no matter what it was we were doing, we were being counted down and expected to execute commands as fast as we possibly could. Everything was on a count: getting dressed, shaving, pissing, shitting, eating, showering...EVERYTHING! This process of being counted down by the DIs was something called "living by the numbers".

Getting dressed in the morning became one of the worst parts of the day as we would take on and off our clothes one piece at a time upwards of 50 times. This was either because recruits were still moving after the DI got to "0", recruits weren't screaming loud enough, or just because the DIs wanted to fuck with us and get us angry. Every morning would start with the dreaded, "LIGHTS, LIGHTS, LIGHTS!!" which would be screamed by the last recruit on Firewatch duty. The lights would snap on and within a second the DIs would be out screaming for us to get out of the rack and on line as they counted us down.

Then there would be a head count, some relentless "fuck-fuck games" and then eventually head calls. A recruit could not simply walk into the head to relieve themselves; they had to wait for the DIs to scream, "Port side shower up, starboard side shave up" and depending on what side of the squad bay a recruit was on (I was on starboard side) they would run into the head and, like a conveyer belt, shave, piss, shit, and then rotate into the shower while port side tended to the toilets and sinks. The DIs were always in there with us too, so we had to try and avoid their detection otherwise we'd be messed with.

Also, we didn't get to leisurely go in there and take our time… NO! The DIs would be counting us down from 60 seconds, running around, and screaming like crazy. They would literally scream, "Shave the left side of your face right now!", and we'd all scream, "AYE SIR!", then shave the left side of our face, and so on until the 60 seconds were done.

After several minutes, everyone would be back on line in front of the racks sporting shower shoes (flip flops) and a towel waiting for the DIs to "dress us" which could happen as many times as they felt necessary. After we were "dressed", we would be instructed to "form up" outside – meaning run and get into formation outside of the barracks in order to be marched to the chow hall.

It would still be pitch black outside and freezing cold. The wind would whip fiercely off the swamps and hit us with a bite that felt akin to an arctic chill, freezing our noses and fingers as we slowly marched to chow. My hands would literally be frozen and numb as they gripped the cold steel of my rifle, which was toted with us all…everywhere we went!

When we would get to the chow hall, we would stack our rifles which would then be watched over by two recruits that the DIs would designate as "late chow" recruits. Some recruit was always watching over something as gear was never left unattended. We would then line up outside the chow hall while getting screamed at. Usually during that time is when we would be screaming back all the Marine Corps knowledge we were needing to know such as "First woman in the Marine Corps", "two Marines two medals?", "treat a burn,", and many more of these "diddies". After a while and after much screaming the DIs would finally let us all inside the chow hall.

Now eating in the chow hall was always welcomed because everyone, including myself, would be starving. The food was not that great and was reminiscent of school or hospital-type food, but we could eat as much as we could fit on our plate. The only caveat to it all was that we had about 3-5 minutes to get it all down, depending on where a recruit stood in the line, and of course there were no going up for seconds. We literally had to shove it all down our faces as quickly as we possibly could and then run back out to formation to continue with the training for the day.

Eating in the chow hall was not a quite time either; we had to sit there, erect, feet at a 45-degree angle, head and eyes straight to the front, and not say a single word. We did not chit chat; we did not converse; we did nothing but eat and scream if a DI zeroed in to harass one of us...which they often did. It was not uncommon for the DIs, especially DI Stubb, to kick a tray of food, spit in the food, or just fuck with us relentlessly while we tried to eat.

The Drill Instructors were on their worst behavior during first phase, employing any trick or tactic in the book to break us down. Sometimes DI SSGT Stubb would find recruits attempting to dispense milk in their cup only then to scream at them and force them to drink cup after cup as punishment, since for whatever reason we were not allowed to drink milk - one of the many ridiculous rules of my platoon.

Sometimes the DIs would walk around screaming and jabbing their fingers into the trays of food and even sometimes spitting in them just for their own amusement. After the morning chow, we would run outside, form up again, and march back to the squad bay to begin the morning clean-up - another terrible time of the day! That is where I was first introduced to the cleaning method, or some may say hazing method, of "scuzzing the deck".

SCUZZ THE DECKS!

Again, unlike the movie "*Full Metal Jacket*" which depicts Joker and Cowboy "swabbing" the deck with mops alone in the head while they converse with each other, we did things a bit different during my tenure. There was not a broom or mop to be found on Parris Island, at least not that I ever saw or used. The way we "swabbed the deck" was something DIs dubbed "scuzzing the deck" and could be achieved either with a scuzz brush, which was essentially a shoe polish brush, or a towel often referred to as a scuzz towel.

There was no leisure time while we "swabbed the decks" either; everything we did was meant to put our bodies in pain. We scuzzed the deck on our faces, in a push-up position and were never allowed to have our knees hit the deck. We would have to run up and down the squad bay floor in a push up position, scuzzing (or moving) out all the bleach and pine sol through the front hatch of the barracks.

That was often accompanied by DIs running behind us and screaming at us for us to scream back as loudly as possible and of course restarting over and over again whenever someone's knee inevitably touched the deck. The act of restarting was called "get back" and could happen as many times as the DI wanted. When they saw a recruit break down and let their knee hit the deck, they would scream, "GOOD, THIS BITCH'S KNEE JUST TOUCHED THE DECK...GOOD! GET BACK RIGHT NOW!"

Scuzzing the deck was one of the most painful activities DIs had recruits perform in bootcamp and one that could most easily break a recruit in just a matter of minutes. There were some recruits that would quit just from scuzzing alone as it was a real morale breaker.

The DIs often chose the recruits they hated the most to do the scuzzing (in addition to being used as a punishment), and I can confidently say that I scuzzed my fair share of decks during my time on Parris Island. Sometimes, if the DI was really in full kill mode, they would make me scuzz the deck with my cover, blouse, seabag, or any number of other items they could think of to have me use. I carry lower back and knee pain with me today as remembrance of scuzzing the deck.

Most of the hazing and "hands on" training occurred in the squad bays where there was no real risk of an officer or very high ranker catching. Basically, if an officer didn't see it then it didn't happen. The DIs were very good at circumventing the various rules and regulations which they were supposed to abide by. For example, it was always said that DIs were not supposed to swear at recruits or put their hands on a recruit and that it was against the rules for them to do so. Well, in the squad bay, if it called for it, it happened and it happened quite a bit.

As the first couple weeks or so continued, several people from the platoon were dropped for either quitting, having some type of melt-down, or for some type of medical reason. During the first few weeks, we had a lot of classes which occurred in a semi-circle shaped building several blocks from the squad bay where my platoon was housed. The classes would be on Marine Corps history, knowledge, customs and courtesies, general orders, and basically everything in between which a future Marine would need to know. These classes were so hard to get through, not because they were particularly difficult or because the information was hard to digest, but because everyone was so tired.

We had to sit erect, feet at 45-degree angle, and keep our eyes wide open. The DIs would patrol around looking for a recruit bobbing for cock (dozing off) and if they found one, which they always did, they would drag them outside and IT (incentive train) them, which included smoking a recruit with all types of physical training and anything else they could think of to break the body, mind, and spirit. It was never good when a DI would scream, "I'm about to break you the fuck off!", meaning they were about to run a recruit's ass until they were half dead. They broke me off on a regular basis during my stay on Parris Island, especially during first phase.

There was a recruit in my platoon whose last name was Shelby, and he seemed to be basically a well-rounded recruit. He didn't look like the type of recruit that would be dropped or would break down. He had no real issues, was physically fit, and seemed to be a good future Marine. To me he seemed like the starting all-

American high school quarterback. Well, I remember at the end of one of the classes for the day, as we were grabbing our rifles and about to run out of the building to form up outside for evening chow, Shelby was still sitting in one of the chairs. I glanced quickly as I was slinging my rifle and saw my kill hat, SSGT Stubb, and several other DIs swarming him with intense ferocity. They were screaming at him, shaking him, pushing him, and trying everything in their power to get him to stand up and comply with their orders. I'll never forget the blank look on his face – it was the look of defeat. His face just screamed, *fuck this shit!* He was completely frozen and would not move. Later that night, Shelby was gone; another one bites the dust.

JUST CAN'T HACK IT!

That's usually how it would happen: a recruit would just freeze, break down, or threaten to kill themselves or a DI and that was it, they were gone. Of course, that was not before some gentle abuse by the DI's. Recruits did not just simply quit on Parris Island; it wasn't that easy. The DIs made good on their promise to "… give every effort to train you, even after you have given up on yourself…" – part of their oath, and a would-be wash out recruit was given all the "incentive" to keep training. That included berating them, punishing them, and then when all else failed, punishing the whole platoon in an attempt to motivate the wash-out recruit to continue training. It worked sometimes, but most often it didn't. Simply put, some recruits just couldn't hack it. If a recruit couldn't make it through the rigors of bootcamp, then they would never make it for four years in the Marines.

De'fuccio was the name of a recruit whom I had met at the

hotel the night before shipping out to Parris Island. We stood outside smoking a cigarette together talking about how motivated we were to make it through bootcamp and become Marines. We were both going to be 0300's and probably would have been assigned to the same unit. I liked De'fuccio. He was a southern boy from Alabama and spoke with a thick southern drawl. I remember him telling me, "I can't see so good out of my right eye"

"So, you lied during MEPs?"

"Yea, but it's okay. I just plan to shoot left-handed!"

I didn't say anything to him and didn't even know if he was being truthful or just trying to be cool but thought that it probably wouldn't turn out well for him. During the first few days after meeting our DIs, Defuccio appeared to be in good spirits. I remember him whispering to me in the chow hall one afternoon while the DI was at the other end of the table, "This isn't so bad. I'm already eating better than I have in years!"

About a day or two later, his spirits tanked for some unknown reason, and he began to malinger. He was placed on light duty for an apparent "foot problem", which meant he couldn't be IT'd or participate in PT or any other event. It was clear that Defuccio was struggling and trying to find a way to quit without saying he actually wanted to quit. A day later when we were in the chow hall again, Defuccio was crying inconsolably and was unable to eat or participate in anything. He was crying so hard and everyone at the table was trying not to stare at him lest they drew in the attention of one of the DIs.

"This is horrible! I've never seen anything like this before", he sobbed as tears fell into his chow tray. About a day later, Defuccio was gone. I never saw him again and don't know what became of him to this day. And that's how it happened. Another one bites the dust!

Sometimes recruits would threaten to kill themselves in a desperate attempt to get off the island and away from the manic DIs. This was NEVER met with compassion and was always treated in the most hostile way. I can remember recruits from different platoons being brought up into our squad bay in the middle of the night accompanied by two or three angry drill instructors. They would run them into the head, stand them in front of the mirror, and make them scream at their own reflection.

They were recruits that had either quit or threatened to kill themselves and they would force them to scream phrases like, "THIS RECRUIT WANTS TO KILL HIMSELF! THIS RECRUIT IS A BITCH! THIS RECRUIT IS A PUSSY! THIS RECRUIT IS A TRAITOR!" and on and on for upwards of an hour. Hearing those screams echo through the squad bay at 2 am was haunting. I couldn't believe how serious the DIs were, and it was clear to me that they were not just making Marines…they were making machines!

KILL RECRUIT VERLICE!

I made it about two weeks before my DIs got my name and therefore had my ass. It all started one night on fire watch. Every night there was a fire watch, and each shift lasted one hour at which point the next recruit on the list, made up by the scribe of the platoon, would come and relieve the current fire watch. Each fire watch consisted of three recruits: the front hatch fire watch, rear hatch fire watch, and then the rover. The duties of the front and rear hatch fire watch were to simply stand their post, which was made up of three stacked foot lockers, and report anything going on at either the front or rear hatches. If anyone came on deck (i.e. Drill Instructors coming into the squad bay after lights out), the fire watch had to report the post and report it correctly.

The rover's duty was to walk – or rove – around the squad bay and make sure all the sea bags, rifles, footlockers, and recruits were secured, locked up, and in their racks. If a drill instructor demanded the rover to report their post, they had to account for all the recruits, sea bags, rifles, footlockers and so forth in a concise manner. On that night I was assigned as the rover – the first shift rover immediately after lights out. It was in fact my first, and only time, being the rover the whole time I was on Parris Island and since it was my first time, I was unsure of what I was supposed to do. I didn't know how to report the post for rover and was overall confused about what exactly my duties as rover were.

After lights out and after the DI went into the DI house, I began walking around and getting the counts. I didn't really have a sense of urgency and was taking my time not realizing that a DI could storm up into the squad bay at any time and demand I report my post. After only about 15 minutes or so the silence of the pitch-black squad bay, which was only illuminated by the glow of the bathroom (head) lights, was shattered by a DI tearing into the front hatch who then quickly made a darting march towards the fire watch. It wasn't just any drill instructor either, it was the chief drill instructor!

The chief drill instructor was SSGT Biggins, and he was one of the most intimidating DIs I had encountered on the Island. The guy had a scream like something straight out of hell – pure nightmare fuel. He was cross-eyed with a face pockmarked full of acne scars. His arms were so big they looked as if they would bust out of his cammie sleeves with even a single clench of his fist. SSGT Biggins came in with one mission: to fuck up the rover. He immediately darted over to me and demanded in one of the most commanding voices I had ever heard, "Bitch, report your post!" I began to studder and stammer, not really knowing what to say, and began to report, incorrectly I may add, my post. "Sir, recruit Verlice reports… ugh…80 foot lockers…"

He cut me off mid-sentence and screamed, "BITCH! I KNOW WE GIVE A PROPER GREETING OF THE DAY! DO IT AGAIN!"

"AYE SIR!", I screamed as I began to start over.

"SIR, GOOD EVENING, SIR. RECRUIT VERLICE REPORTS 80 FOOT LOCKERS, 80 SEABAGS, 80 RIFLES, AND 80 RECRUITS ALL LOCKED AND SECURED! GOOD EVENING, SIR…" He then screamed back, "BITCH STILL WRONG!" He was obviously getting more and more angry each time I messed up the post reporting and I could tell that he was going to make me continue attempting to report the post over and over again. "BITCH!", he yelled, "WHAT HAVE YOU BEEN DOING FOR THE LAST 30 MINUTES? YOU SHOULD ALREADY HAVE THESE COUNTED!" I didn't know what to say to him, so I just screamed, "AYE SIR!"

Angered and feeling as if though I had defied his every command, he began to play what they call "fuck-fuck" games with me. He started making me run around the whole squad bay counting every footlocker, sea bag, recruit, rifle, etc.... over and over and OVER again. After roughly 30-40 iterations of that, I let out a huff. Looking back on it, it was sort of like a huff of irritation or annoyance...big mistake on my part! SSGT. Biggins took my huff as the ultimate form of disrespect and began to chew me out loudly for every recruit laying in their rack to hear. It was really the first ass chewing that I had received directly one-on-one from a drill instructor since arriving at Parris Island, so I didn't know how to respond to him other than screaming back, "AYE SIR!"

He screamed and berated me for a while before explaining to me why recruits do fire watch and lecturing me on the right way to get a count. He didn't put his hands on me or anything and essentially just left it as a good old-fashioned ass-chewing. Little did I know, that was going to spark a chain of events which pinned me as a problem recruit for the rest of the time I was on Parris Island. SSGT Biggins left the squad bay through the front hatch from which he had originally entered, and I continued to rove around the squad bay feeling as though the worst of it was behind me. I even remember snickering to some of the other recruits who had witnessed the event and cracked jokes about how he was being an ass.

Several minutes later, as I continued to rove around, the door to the DI house whipped open, allowing a bright yellow light to bleed into the dark squad bay. My eyes shot to the door where all

I could see was a dark silhouette of a body clad with a smokey the bear cover. I felt my heart sink.

"WHERE IS VERLICE?!", barked the dark figure. The figure started to march forward into the squadbay with purpose, stiff and rigid with the sharpest stride I had ever seen. As he got closer, I could see that the dark figure was my heavy hat, SSGT McHolland. I could see that his brow was creased in anger and his lips were curled downwards with a frothy mix of anger and homicidal rage as he laid his soul piercing eyes on mine.

"Here Sir...", I managed to muster with scant amounts of courage. "GOOD! You're already up!", he shouted as he made the final leap towards me. His speed and strength shocked me and within a moment he was on top of me, his scrawny hand clenched around my neck like a noose. He probably only weighed 160 pounds or so but was able to pick me up and heave me 20 ft backwards like I was a ragdoll.

I smashed into the rear hatch Firewatch post with tremendous force, causing the three stacked footlockers to tumble down as I struggled to catch my balance. Before I could straighten myself up, SSGT McHolland was back on top of me, grabbing me by the throat and peppering me with punches to the stomach and ribs. It was clear to me in an instant that the guy meant absolute business.

SSGT McHolland took off his duty belt and campaign cover and haphazardly threw them on the floor at his feet, which was significant as DIs did not usually treat their own gear that way. He grabbed me by the collar and yanked me close in to where his nose was practically touching mine. What he said to me has never left my memory.

"I don't give a FUCK who you think you are, where you're from, or what you did before coming here! I will fucking KILL YOU! I have killed more people than there are in this squad bay, and I promise you, bitch, I will kill you before you can graduate here!" His eyes almost seemed to be glowing red, and I could feel them burning the last shred of dignity I had left as he continued reprimanding me. "from here on out, every morning and every day, you will be on that quarterdeck! You will be fucked up every single day until you either die or quit! You will NOT be a Marine as long as I'm here, SCUMBAG!"

He finally let me go before picking up his cover and belt, which he placed back on his body, and then he disappeared back into the DI hut. There I stood alone in the cold and dark squadbay in a total state of pure trepidation. I knew from that day forward I was finished. I had been targeted; they had my name, and I could only imagine what was going to happen to me in the coming days and weeks. I began to wonder if I would indeed make it through bootcamp or if I too would end up like Defuccio.

The next morning upon lights on, sure as shit, SSGT McHolland came running out of the DI hut and immediately came over to me, grabbed me, and dragged me to the quarter deck where he then began to smoke my ass with all types of pushups, side straddle hops, mountain climbers, plants, sit ups, and everything in between. This quarterdeck treatment occurred every day and whenever there was a moment, I was pulled to a sand pit or quarterdeck to be IT'd. The punishments were not limited to physical exercise only; he also mentally punished me as well with all types of taunts and other tactics to break me down.

A couple days later after the Firewatch incident, SSGT McHolland refused to let me go into the chow hall for morning chow, which was something the DIs were NOT supposed to do. Later that day, he again wouldn't let me go into the chow hall for lunch. It was a tactic being used to break me and force me to quit or perhaps to make me pass out so that I would be dropped for medical reasons. I was not going to let that happen and later that day when we were cleaning our rifles in the squad bay, I requested to speak to the Senior Drill Instructor and was ultimately granted permission by SSGT Stubb. I banged loudly on the hatch three times and screamed for permission to enter the DI hut and finally SSGT Fierro let me in where I tried to then gingerly tell him that one of the DIs would not let me eat.

He seemed shocked and demanded that I tell him which one was responsible. I hesitated, because I did not want to look like a snitch, but ended up telling him it was SSGT McHolland who was responsible. Luckily, SSGT Fierro was a fair senior DI, and he simply nodded and allowed me to eat a box chow in the DI Hut while he watched me closely. That was essentially what the Senior DI was there for – for that exact type of grievance. If I had not requested to speak to him about the chow hall incident, I most likely would have passed out at some point from exhaustion and lack of food.

I ate as quickly as I possibly could since I was feeling very uncomfortable being watched over by the senior DI. I remember he looked at me as I was shoveling food into my face and said, "Verlice, do you really want to be a Marine?" I responded without hesitation, "Yes, Sir!" He nodded and said, "not everyone makes it you know." I felt like he was trying to tell me something but tried not to let it get into my head and just shrugged it off as another one of their many mind games. As I finished the box of food, I glanced over to my right where my eyes caught the whiteboard which the DIs wrote their plans for the day as well as other notes which they deemed to be important. In the middle of the board, written in large bold words circled three times with arrows pointing to it, read:

KILL RCT VERLICE!

CHRISTMAS DAY

Christmas came about three weeks after I had arrived on Parris Island and I remember wondering if we would be allowed some extra free time (we got about one hour of free time a night on most nights – where we could write letters, shit, shower, shave etc. without DIs messing with us…too much). I remember thinking of movies like *Full Metal Jacket* which depicted GSGT Hartman singing a corny Happy Birthday, Jesus and telling the recruits that there will be a "magic show…" later. Well, that did NOT happen on Christmas Day in Mike Company, platoon 3080. It was almost as if the DIs were trying to make Christmas day as horrible as they possibly could by inflicting as much mental and physical pain as they deemed necessary. Christmas Day on Parris Island was the day of the "Scuzz Rally 500" - the ultimate haze-fest!

I remember all the DIs, including SSGT McHolland, SSGT Stubb, and our new drill instructor who was fresh out of DI school and in his first rotation, SSGT Smith, got the recruits in the squad bay to make the house "float". Making the house "float" was a painful process which involved every recruit picking everything that wasn't bolted onto the ground (ie. footlockers, racks, seabags etc.) up off the deck making it "float", all while screaming at the top of our lungs. The DIs would scream and make us scream back and if someone dropped something or if something touched the deck, we would be ordered to put all the stuff into one corner of the squad bay so that essentially there was a wide-open space all throughout. It was a type of fuck-fuck game where the DIs would order us to do impossible things that could not be achieved and each time we failed, the orders got worse and more painful.

Before long, we were running mattresses into the rain room (shower room in the head), stacking footlockers in one corner of the squad bay and the racks in the other corner. All the extra gear was then haphazardly thrown into the opposite corner of the squad bay onto the quarterdeck. After screaming and running around all over the place, the DIs ordered us to pull out our scuzz brushes and get on our faces. DI SSGT Stubb screamed, "ALL YOU BITCHES WILL SCUZZ THIS DECK IN A PUSH UP POSITION IN A CONTINUOUS CIRCLE AROUND THE SQUAD BAY, LIKE A RACETRACK!"

"AYE SIR!", we all screamed.

"IF ONE RECRUIT'S KNEE TOUCHES THE DECK, WE ARE GOING TO CONTINUE FUCKING AROUND ALL DAY LONG, DO YOU UNDERSTAND?!"

"AYE SIR!"

"LET ME SEE A BITCH TAKING A BREAK! LET ME

SEE A BITCH NOT PUTTING OUT AND I PROMISE YOU, YOU WILL BE IN MORE PAIN THAT YOU ARE ALREADY GOING TO BE IN!", he screamed ferociously. "AYE SIR!", the whole platoon screamed one final time as the games begun.

All 78 recruits began to scuzz the deck, in push up position, around and around the large squad bay. After only one round about the squad bay, the knees started to buckle, the thighs started to burn, and knuckles got rubbed down to nothing. The event may seem like it had a purpose - like perhaps it was to build confidence or get future Marines ready for the rigors of combat? In all reality, it was done for no other reason than to just physically break us down; and break us down it did. The "Scuzz Rally 500" carried on for two to three hours and was one of the most painful moments I had had up to that point in my life.

As the "game" continued, and the screaming rose to another level, tensions began to rise amongst the ranks. These games, in addition to causing pain, caused anger among many of the recruits. This was also part of the MO of these fuck-fuck games – to anger us. After a couple hours, recruits began to fight with each other, push each other, and scream obscenities at each other as we all began to reach our breaking point.

One of the squad leaders, recruit Robertson, pushed me and I had had enough! I rose to my feet as he was rising to his and before I knew it, we were in each other's faces about to start swinging on one another. That sort of tension was new for me as I had never been in a fistfight my entire life. It shocked me how much anger I was feeling towards him. The Marine Corps was working on me already!

The would-be fight was broken up by another recruit and for some reason I was grabbed by SSGT Stubb who then began to berate me and scream at me. I was pulled one way and then another way by two of my DIs and then blasted with spit as I was continuously screamed at for several ignominious moments. Two to three hours of the "scuzz-rally 500" truly had me at my breaking point and ready to snap, but I just screamed, "AYE SIR!" to the DIs and then continued with the event.

Shortly thereafter, the DIs made us get back on-line at the position of attention. The event was over and everyone was broken, angry, and discouraged. I remember distinctly my lower lip quivering uncontrollably, but I did not cry. It was probably the closest I came to crying while at bootcamp (outside of the EGA ceremony that is). That event just let me know what levels the DIs were willing to take things to bring the recruits to physical pain and their mental breaking point. The rest of the day ensued with typical first phase bootcamp activities and that was my Christmas of 2011.

SSGT SMITH

Like I mentioned earlier, SSGT Smith was a new drill instructor who had just graduated from DI School and was placed into our platoon, as a second kill hat, around the second or third week of bootcamp. He was what was considered a "boot" DI meaning he was brand new. The new DIs like SSGT Smith were probably the easiest DIs to deal with simply because they punished us strictly by the rules of the DI school and the rules of the Marine Corps. The Marine Corps was strictly against hazing, swearing, abuse, and any other illegal used tactics to break either recruits or Marines and trained their DIs to adhere to such standards. Hazing was a zero-tolerance activity - no different than drug use.

The issue was that the Marine Corps depends on these hazing and hands-on abuse tactics to make Marines – like the Marines who raised the flag on Iwo Jima in 1945 and the Marines who fought with their dicks frozen off in the Frozen Chosin Reservoir during the Korean War. This sort of animalistic and machine-like approach to fighting war is an expectation we were expected to live up to as a Marine and the only way to make us that hard was to train us that hard. The boot DIs, like SSGT Smith, played strictly by the rules but quickly learned from more hardened DIs, such as SSGT Stubb and McHolland, that there were ways to circumnavigate these rules and still give recruits the training required to become a Marine. Eventually, they all turned hard like SSGT Stubb; it was just a matter of time.

SSGT Stubb was certainly one of the hardest drill instructors on Parris Island at the time. It was said among recruits that his screams could be heard on the opposite side of the Island, and I firmly believe that was true. He was also incredibly sadistic even for a kill hat. He took pleasure in breaking recruits down and making them do disturbing or macabre things just either in the name of Marine Corps Recruit Training or for his own entertainment. I can recall one moment during first phase, while we were all sitting on the quarter deck, some kid who was wearing a blue mask and obviously sick, began to vomit copiously. DI SSGT Stubb ran over to him and made him take the mask off and put it back on multiple times essentially forcing the kid to eat his own vomit. "NOBODY WANTS THAT SHIT, BITCH!", I remember hearing him scream as the recruit gurgled back, "AYE, SIR!"

Another incident pertaining to SGGT Stub which I recall was when a small-statured recruit from another platoon was brought up to our squad bay to be harassed by several of my platoon's DIs. It would happen sometimes where some random kid was brought up onto our deck for unkown reasons simply to get hazed by a different set of DIs. The small kid was being smoked by several DIs and as a result was crying inconsolably. The moment SSGT Stubb saw the recruit crying, he approached him, grabbed his face to wipe his tears, and then snorted them before screaming, "BITCH! NOW I HAVE YOUR SOUL!" I remember thinking to myself, *did he really just do that?*

Yes…yes in fact he really did do that! He knew all the tricks in the book to break recruits mentally and physically, and he knew every trick to get around the "no hazing laws" which the Marine Corps had in place during recruit training.

Anyway, regarding SSGT Smith…

One morning during first phase, immediately after lights on, SSGT Smith randomly grabbed me during the morning clean up and took me outside to the sandpit to smoke me. It was sort of unprompted and looking back in hindsight, I think he was just trying to sharpen his IT skills on recruits. Every time we either got on the quarterdeck or a sandpit to get smoked by a DI, we immediately stripped our blouse so that we were just in uniform trousers and the green skivvy shirt.

I was getting smoked by SSGT Smith for about thirty minutes or so – way too long for the morning – and once he realized that he, screamed at me to run back up the three flights of stairs to the squad bay. I complied and once I got into the squad bay, I saw that all the recruits were already lined up with SSGT McHolland counting everyone down. He was already on 11 by the time I began making my way to my spot on line and I knew that if he caught me after he reached "ZERO", I would be dead meat. I had 11 seconds to get there with my blouse back on and at the position of attention.

The race was on and I ran as fast as I could, quickly throwing my blouse back on as fast as I possibly could to my spot on line. "7, 8, 6..." the count continued as I sprinted desperately. "4, 3, 2..."

I had gotten to my spot and got to the position of attention, but as SSGT McHolland screamed "ZERO!", indicating that all moving from recruits stopped, I had not finished getting my blouse on; there simply just was not enough time for me to do it.

SSGT McHolland immediately saw that I was not properly dressed and ran over to me in a rage. "MOTHER FUCKER!", he screamed, getting the entire squad bay's attention. "SO VERLICE JUST WANTS TO DO WHATEVER HE WANTS TO DO AGAIN RIGHT?"

"NO SIR!"

"BULLSHIT, BITCH. YOU DON'T GIVE A FUCK ABOUT ANYONE BUT YOURSELF, RIGHT?"

"NO SIR!!", I screamed again as loudly as I could.

Unfortunately, I couldn't simply tell SSGT McHolland that I was outside getting smoked for no reason by SSGT Smith – it didn't work that way in bootcamp. I couldn't reason with the guy and explain my situation in any way. I had no choice but to take whatever punishment was going to befall me.

"GOOD, MOTHER FUCKER!", SSGT McHolland bellowed, "SQAT!"

"AYE SIR!" I screamed, getting into a squat position, ready to take the punishment that I knew was going to break me off.

SSGT Smith ended up taking the platoon to the chow hall for morning chow while I stayed in the squad bay with SSGT McHolland getting that good one on one attention. My punishment consisted of an elevated push up position –feet propped up on my footlocker – one arm in the air holding my rifle by the front sight assembly with my pinky. It was like a modified one-handed elevated push up all while holding a rifle.

It was PAINFUL too!

I was in that position for roughly 30 minutes before the platoon came back from chow. I was beginning to lose my strength and could barely keep myself up especially with the occasional kicks from SSGT McHolland. Sweat pooled on the deck beneath my head as my strength to hold myself up became more of a struggle. SSGT McHolland was kneeling next to me the entire time screaming and spitting, "GET BACK UP NOW VERLICE!" every time I started to dip down.

"AYE SIR!", I managed to get out each time, just barely able to lift myself back up. At one point during the punishment, SSGT McHolland got close to my face and spoke in a hushed but firm tone, "I fucking hate you, Verlice! And just know, if you don't quit today, I will keep doing this to you every single moment of the day until you do!"

It was clear that DI SSGT McHolland was going to see to it that I was dropped and was going to do everything in his power to ensure that he made that happen.

Eventually, I was told to get up and out of the position I was in only to be quickly whisked out of the squad bay by SSGT Stubb. Once out of the squadbay and into the stairwell, SSGT Stubb made me run up and down the three flights of stairs over and over again all while counting me down from ten. I had ran up the stairs so many times that I thought I was going to faint. Eventually, he ran me into a different platoon's squad bay where there were two or three other recruits on the deck being hammered by their platoon's respective DIs. It was there that I had another one of the many terrible experiences I endured while at bootcamp.

SSGT CORDOBA

One of perhaps the meanest and cruelest DIs I ever came across while I was at Parris Island was this other platoon's senior DI. Thankfully he was not one of my DIs otherwise I may not have made it through bootcamp because the guy was pure demon. After being brought up onto the squad bay, I endured about 40 minutes of abuse which included getting screamed and sworn at by 5 different DIs at the same time, scuzzing the deck with my cover, and then getting that cover forced into my mouth by several of the DIs. "EAT THE CHOW, BITCH!", they mocked as they forced it as far into my face as they could get it. It was after those events that I was approached by one of the sickest DIs I would end up meeting on Parris Island.

SSGT Cordoba, a tall, lanky, pale and sweaty man, stood in front of me and pulled me in close by the collar so that his face was directly in mine. He bellowed to me in an impishly calm voice, "Mother fucker, do you know why this is happening right now?" I remember feeling fear grip me as his satanic words rang around in my ears.

"You better scream every time you walk by me!" He stared at me with a look of homicidal rage as he continued, "The other morning after haircuts as you were running back to your platoon, you ran by me and didn't give a proper greeting of the day!" I noticed the man had black lifeless eyes; his thin, cracked lips were as pale as the rest of his wet, clamy skin. "I will haunt your fucking nightmares here! I will find and rape your mother, sister, father, brother, and I will rape you too, bitch!" I couldn't believe what I was hearing from the guy and by the veracity in the tone of his voice, it sounded like he meant it. All that because I had failed to give him a proper greeting days before?

"NOW SCREAM!", he blasted in my ear, covering the side of my face with hot spit. I screamed back as loud as I could, "AYE SIR!", and was then grabbed by SSGT Stubb and pulled back to my platoon to continue the day. The whole event lasted about an hour and left me shaking for the rest of the day. I laid in my rack that night having nightmares of the evil senior DI and the things he said to me that day.

The words of SSGT Cordoba, although purely just to scare

me and break me mentally (hopfully) have remained with me to this day and I can still feel his hot spit sliding down the side of my face. I remember my mother had told me before I had left home, "just remember, it's their job; don't take it personally." Her words echoed in my head but did not resonate with me over that encounter.

It was very clear to me that things were personal to the DIs; making Marines – like the type who raised the flag on Iwo Jima – were all the DIs cared about. The DIs did not want to see any recruits falling through the cracks. If a recruit didn't deserve to serve in the Corps, they were going to try their hardest to do anything in their power to break that recruit and wash them out. It didn't work on me, but they tried.

SWIM QUAL

Towards the end of first phase, the recruits were required to do swimming qualifications. Swimming qualification, or swim qual as we called it, was an event where the recruits were marched to the base pool by the DIs and put through several swimming events/exercises including breaststroke maneuvers, back stroke maneuvers, swimming with a full gear load, being submerged in water and stripping off all the gear, and then finally jumping off the platform in flak and Kevlar and keeping ourselves afloat using our blouse tops as a makeshift life preserver. Swim qual was a pass/fail event and if a recruit failed it on their first attempt, they were given one more attempt and if they failed again, they were automatically dropped regardless of overall performance in bootcamp. The recruits that failed a swim qualification were labeled "iron ducks".

I was very nervous about the event because I was not the strongest swimmer and while I could keep myself afloat in water and swim well enough, I was not sure about doing it all with a full gear load. Being marched into the pool facility caused heart palpitations and great apprehension as I was worried something would happen to where I would be dropped. I felt like I had been through the ringer in only four weeks of bootcamp and still had 9 weeks left. My sights were set on becoming a Marine and I wasn't about to let a swim qual get in the way of that.

The swim qual was led by USMC swim coaches who were not DIs and therefore were a bit softer on the recruits, which was a welcome change given the torment of the last four weeks. The first three events, which included the submerge, treading water, and floating with uniform makeshift lifejacket, I passed easily; however, I failed the distance backstroke swim portion mainly because I began to panic halfway through. The failure was a crushing blow for me and ramped anxiety up to the next level. I would have to come back the next day with the other recruits who didn't pass the first time and try again. If I failed the second attempt, I would be labeled an "Iron Duck" and dropped from my training platoon.

The rest of that day I was so worried that I would not pass the second attempt of qualification and prayed to God quite a bit to help see me through it. SSGT McHolland let me know how much he was hoping for me to fail because, again, it was his sole purpose to drop me for one reason or another. He berated me for the rest of the day, taunting me and making crash jokes about how happy he would be once I was dropped. It was not "motivation" being given to boost me but just the opposite: the guy truly wanted me to fail.

The next day I marched back to the pool facility with a handful of recruits who also did not pass the swim qual on the first day, and while I was incredibly nervous, I passed the backstroke event which I had failed the day before and passed the rest of the swim qual. It was like a huge weight lifted off my shoulders and I felt that I could possibly get through the rest of bootcamp even with SSGT McHolland on my ass 24/7. I still remember on that day it had been a woman instructor with the pool who coached me across during the back stroke portion of the qualification. She could tell I was nervous because of my breathing, and I really felt like it was her voice and calming presence that helped me across the pool. I always seemed to do better in the presence of women but unfortunately that was the only time I ever dealt with a woman Marine the whole time during my military service.

Later that day, feeling much lighter and with no threat of being dropped, we all prepared for the final two events of first phase: the rappel tower and the dreaded gas chamber! I was probably more worried about the gas chamber than anything else during bootcamp, and I was not looking forward to being sardine-packed in a room filled with gas. I went to sleep in my rack that night feeling as though the title of US Marine was finally in my sight.

THE GAS CHAMBER

Early in the morning on the day of the gas chamber and rappel tower, the recruits were marched to the area where both the gas chamber and rappel tower were located. I remember looking at the gas chamber thinking that it looked like one of those rooms that Nazis would march innocent Jews into to be exterminated. I tried not to think of morbid thoughts as we all listened to the directions on how to operate the gas mask and what to expect in the gas chamber.

They explained to us that we would march in the gas chamber and begin a series of exercises, such as jumping jacks, pushups, and running in place just so that we could see how the mask would stay on our face and keep the gas out – essentially showing us that the gas mask works and that we were to trust in the mask. Then they instructed us that we would have to "break the seal", which basically just meant we would lift the mask up off our face so that the gas would enter. Then it was our job to remove the gas from the mask by holding the canisters and exhaling out hard so that the gas would "blow" out of the mask. The exercise was done to show us that if we were in a real poison gas situation and some of it had entered the mask, it was possible to remove it and continue with the mission.

The maneuver does work…partially; but once the gas hit my face and the searing pain gripped my eye sockets, I inhaled deeply, which immediately caused a heaving and coughing fit as well as a great sense of panic. The panic makes clearing the mask much harder. The taste of the gas was strong and virulent to the eyes, nose, and throat and as moments went on, all I wanted to do was get out of the chamber.

After what seemed like hours, they opened the door to the chamber and we filed out, ripping our masks off and coughing, gagging, and hacking. Every recruit was leaking snot like a faucet from every orifice on their face. Some recruits even puked from the intense hacking the gas induced. I began to feel relief that it was over and that I had made it through …or so I thought!

As I rounded off, still coughing, SSGT McHolland grabbed me by the collar and threw me back in line with a bunch of recruits who had still yet to go into the gas chamber. "Bitch, you're going in again!"

My heart sank! I felt like I barely kept my shit together inside the chamber the first time; how was I going to make it the second time? That was the first and only time I began to beg the DI for mercy. "Please, Sir! Don't make *this* recruit go back in!" All the dignity drained from my body as the words escaped my cracked lips.

"BITCH!", SSGT McHolland barked, "YOU'RE GOING BACK IN!"

"AYE SIR", I screamed back after a moment of hesitation. I was unable to beg any further – I mean, it was pointless. If a DI told a recruit to do something, then damn it, they're going to do it.

A couple of other DIs which were in the area began to torment and taunt me as my fear was clearly palpable. One of the DIs grabbed my mask and began to rip the filters out of the canister! Fear shot through my body at the sight of that because I knew without the filters, the mask would not work. I would be in that chamber for the second time, the whole time without a functioning gas mask. That would be very rough for anyone to endure and would be like going in without a mask on at all! I felt like pleading with the DIs to put the filters back in, but it would have just led to more torment and God knows what else. There was nothing I could do, so I just sacked up.

They threw the mask back at me, minus the filters, and I was left with no choice but to go in without a functioning gas mask. I'd like to think that I was a courageous person back then, but I would be lying if I said I wasn't horrified to go back in the chamber without a working mask. I was the only recruit that I was aware of who was forced to go in twice and I was certainly the only one to go in without a functioning gas mask. When I emerged from the gas chamber for the second time, unable to breathe and with tears and boogers flowing from every hole in my face, SSGT McHolland stood close by just shaking his head and walking away. He was trying everything in his power to make me quit and he was not done yet!

The next event which occurred immediately after the gas chamber was the rappel tower. Many recruits feared the rappel tower because of the height and just overall intimidating nature of leaning back and running down a 75-foot wall. I did not really fear the rappel tower; however, I was worried about tying the rope around myself correctly. They instructed us on how to tie the rappel rope around our groin and showed us how to rappel down the tower by leaning back and easing up on the rope to allow decent. After thirty minutes or so, I found myself at the top of the tower ready to lean back and rappel down. I had a brief moment of hesitation as I leaned back mainly because I had to trust the rappel rope which I tied myself, but I leaned back all the way and began my decent.

It was going pretty well until about halfway down the tower, I felt tension on the other end of the line. I turned my head and saw

SSGT McHolland standing below me, holding the bottom half of the rope. He heaved and swung the rope out which caused me to lose my footing on the wall and swing out wildly only to then come crashing back into the tower. He did these maneuvers three or four more times until I reached the bottom. It was very frightening as I had essentially been flung off the tower and thrown back into it several times. If my rope had not been tied correctly, I would have certainly fallen 50 feet to the ground and probably had been injured or worse. Again, I was the only recruit that I knew of who that had happened to. Two times in one day SSGT McHolland had tried to make me quit and still, I stood strong.

Once we finished those two events, we marched to the chow hall for lunch and then back to the barracks where we were given a second set of cammies (uniforms). Those cammies had our names as well as US MARINES stitched onto them. No longer did we have to wear the laminated name tag that labeled us as new recruits. We were also allowed to blouse our trousers over our boots instead of cuffing them. That brought us all a sense of pride, like we were one step closer to looking like Marines and one step closer to becoming a Marine. Now, just because we had the US Marines stitched onto our new uniforms did not mean we were Marines yet...NO! We still had a long way to go before we earned that title.

2C – PHASE II (WEEKS 4-8)

Phase II began early in the morning after the gas chamber and rappel tower and started with a hump to the rifle range barracks which was roughly 7 miles. One thing about the humps in the Marine Corps was they were not leisure walks - they were practically runs! They were designed to break us physically and they almost always did. The moment we got to the rifle range barracks and got settled, Recruit Backer was dropped. It was astonishing that the kid even got as far as he did. Most often the weak recruits were dropped during first phase, and it was very rare for anyone to be dropped in second phase, although some recruits were under certain circumstances. Backer was a weak recruit who just couldn't hack it.

It may sound cruel, but the Marines are built on a foundation of never quitting and never giving up. Throughout the history of the Marine Corps, there has been a demonstration of this tenacity: Belleau Wood during WWI, the islands in the pacific during WWII, the frozen Chosin Reservoir during the Korean War where it was said Marines were frozen from the waist down and still came out victorious, Vietnam, Fallujah, Iraq, and on it goes. The various "fuck-fuck" games that the DIs played with the recruits were designed to test this never quit and never give up mentality. It didn't necessarily matter how fast one was or how accurate they were; if they never quit or give up, then in the eyes of the DIs and the eyes of the Marine Corps, they are worthy of calling themselves a United States Marine.

The Marines have always instilled a fight until the last breath mentality. There was no surrender – it was death before dishonor! That is what differentiates the USMC from the Army. Marines have been trained to fight until they die, never take a prisoner, and NEVER leave a Marine behind. Backer did not exemplify these traits as he was clearly a quitter and therefore would never be a Marine. Backer was gone that day and SSGT McHolland couldn't be happier. He ran up and down the rifle range barrack squad bay screaming, "NAH, NAH, NAH, NAH, HEY, HEY, HEY BACKER GOODBYE!"

The first week on the rifle range consisted of what was called "grass week" where we basically dry fired our M16-A4 service rifle in various positions such as the prone, kneeling, sitting, and standing. That occurred for 5-6 hours per day and became very painful. Try being in a kneel position while staring down the sights of a rifle for 5 hours straight – again all while being screamed at. One kid, Rodriguez, literally got nerve damage on his right leg and was medically dropped, although he later did become a Marine. It was during grass week that we were introduced to a new form of abuse – one that I still shudder about to this day.

BOX CHOW SLUSHIE

Remember the box chows or bag-o-nasties I was talking about earlier on? The box of shit food with the sub sandwich, generic chips, and cold egg? Well, those became a staple of grass week as we did not go to the chow hall for noon-time chow while on the range. During bootcamp, with all the exercise and stress, we became so hungry all the time that the box chows became an amazing treat – a stark contrast from the first day on the Island when they almost made me vomit. I got to the point where I would devour the sub sandwich, chips, and even the egg like it was a five-star gourmet meal – we all did.

One day on the rifle range, during grass week, the DIs gathered us to sit crisscross in formation and then proceeded to have recruits pass out the box chows. With our stomachs rumbling, we began to tear open the boxes and remove the contents for consumption. It was at that moment that SSGT Stubb came tearing through and screamed at us, "STOP RIGHT NOW!"

We all screamed in unison, "AYE SIR!"

He began to give us ridiculous orders such as pulling out the box of raisins and removing one single raisin, holding it in the air, and making everyone count off the raisin one by one. 72 recruits counted one by one, "ONE, TWO, THREE…" and so on until the last recruit counted off, "72 RAISINS ALL ACCOUNTED FOR, SIR!" "GOOD, NOW THROW THE RAISIN BEHIND YOU RIGHT NOW!" We all did as we were instructed and then repeated the ridiculous act several more times.

It was very clear that DI SSGT Stub was screwing around with us by purposely not allowing us to eat knowing full well we were all starving. He then began to order us to take apart the sub sandwich, crush the chips, the egg, and dump the rest of the raisins in the box. As we did that, he then screamed at us to eat the mayonnaise by itself and gave us a countdown of ten seconds to do it. I did not eat it but pretended to so that I would not be singled out. He then instructed us to pour water from our canteens into the box of crushed food and mix it together.

Mixing the water in created a disgusting slush of food that not even a starving kid in Africa would want to touch. The DI then proceeded to order us to drink the slushy of mixed food, and we had twenty seconds to do it. Once again, I did not drink the concoction but pretended to. Every recruit in the platoon was angry because our chowtime had been screwed with. The "box chow slushie" was just another tactic used to make us angry and mess with us mentally. We went without food for the rest of the day until evening chow.

The "box-chow slushie" happened two more times during bootcamp and each time it caused people to freak out and fight with each other. None of us wanted to starve until evening chow, but SSGT Stubb didn't care. He just watched us in bliss as we began to break apart and eat each other alive. Part of being a US Marine is going hungry sometimes – it's just the name of the game.

POISON GAS!

One of the worst events that I endured at bootcamp happened halfway through the rifle range once we completed grass week and moved onto firing week. During firing week, we shot our rifles at targets up to 500 yards away and had to qualify by shooting a certain number of points. Anyone who didn't qualify by shooting X number of points was dropped. Half of the day was spent in what we call the "pits" pulling targets for another platoon; the latter part of the day is when my platoon would shoot while the other platoon would be pulling the targets.

Every morning would start the same with head calls, march to chow, and then back into the barracks while it was still dark outside to perform a super thorough cleaning of the squad bay. During first phase I had been designated one of the "scuzz" recruits, but by the time we were on the rifle range I had sort of been promoted to making the racks in addition to checking them to ensure that they were tight and up to the DIs standards.

Making the racks was much less taxing on the body than

scuzzing the decks was, and I preferred it 100 times over. The DIs by that point sort of seemed to be pulling off me a bit and were not up my ass 24/7 like they had been during first phase, and that made me feel as if though I was growing as a recruit and future Marine. Unfortunately, it was all about to change again one random morning on the rifle range.

During the morning clean up throughout the second phase, the DI on duty would sometimes be in the DI hut so that the recruits could clean without harassment, otherwise nothing would get done. One morning, after I had finished making all the racks and checking them to make sure they were tight, I decided to deviate from the normal which would have been simply standing on line waiting for the DIs to come out of the hut to carry out the plan of the day. I had noticed that there were still a few recruits scuzzing the deck, wiping things down, and doing other cleaning duties as per usual. While some recruits were beginning to stand on line at the position of attention waiting, I figured that I had a few moments to run into the head to fill up my empty canteen. I didn't really see harm in it considering I had finished with my duties.

I had just entered the head and was standing in front of the sink beginning to fill up my canteen when I noticed that everything had gotten eerily quiet out in the squad bay. It was sort of like the way a forest gets eerily quiet when there is a predator lurking around looking for its next meal. I looked up and into the mirror as I capped off my canteen and behind me stood SSGT McHolland glaring right back at me with a rageful grimace frozen on his face.

Where he came from so fast, I had no idea, but he made great haste in grabbing me by the back of the neck, lifting me up off the deck, and then walking me out to the platoon which by that point had quieted as everyone stood on line. I had been in the head for only 30 seconds and was incredulous at the fact that everything had unfolded so quickly. I had been positive that I would have had enough time to fill my canteen and be back out to my spot on line before the DIs came out of the DI hut, but I guess I had misjudged and was about to pay one hell of a price for it.

SSGT McHolland had no idea that I had already finished making the racks and was just filling my canteen before getting back on line, and of course I was unable to explain it to him as recruits did not speak to DI's in that manner. He began to scream to the whole platoon who just watched in silence, "THIS BITCH WANTS TO FILL UP HIS CANTEEN WHILE EVERYONE ELSE DOES WHAT THEY ARE SUPPOSED TO BE DOING! THIS BITCH ONLY CARES ABOUT HIMSELF AND NO ONE ELSE!",which was not true, but seeing that I was just a lowly recruit, all I could do was deal with the wrath of SSGT McHolland.

At some point during his hate speech, SSGT Stubb emerged from the DI hut and SSGT McHolland instructed him to "Fuck this Bitch up!" while he marched the rest of the platoon out to the rifle range for the day. That was bad news for me because it meant that I would be left alone in the squad bay with one of the meanest DIs on Parris Island doing God knows what with no mercy. The senior DI was nowhere to be found that morning, and he was the only one who could put a stop to things if they got out of hand. I was truly going to be at the mercy of SSGT Stubb. I was terrified to say the least, but also ready for whatever he was going to throw at me.

Once the squad bay was completely empty, SSGT Stubb started his games with me. He began by making me get on my face and scream as loud as I could. And then it was the, "UP, DOWN, UP, DOWN, UP DOWN", game while he screamed at me, and I screamed back. Every time I would scream back at him, he would demand that it wasn't loud enough, and I was to get back on my face. Once back on my face, I was at once ordered back on my feet. That game continued on and on for quite some time.

Eventually things amped up and I was forced into the head where a potent concoction of chemicals was mixed in a large yellow bucket. I wasn't too sure what liquids were used but looking back, I'm pretty sure it consisted of some mixture of bleach, pine sol, possibly some other cleaners, and… Aqua Velva aftershave! Why would he put Aqua Velva into a mix of cleaning solution? Well, it was to create whatever poison gas which quickly began to form after the deathly brew of liquid had been combined. Like some type of science experiment straight from the depths of hell, I was the student and SSGT Stubb the teacher. The solution was so strong that my eyes began to burn immediately; my throat began to itch and burn and felt as if it would close at any moment.

Next thing I knew, he had dumped the entire five-gallon bucket onto the head floor and ordered me to scuzz it – on my face in a push up position – all the way out the front hatch of the squad bay. That was not all, however, as he was in full DI kill mode and was screaming, running, spitting, hitting, kicking, and taunting the entire time. I could not move fast enough or scream loud enough, and it became apparent after several moments of attempting to get that liquid out of the front hatch that he was not going to let me complete the task. He was counting me down and every time I would get close to the hatch, he would go straight to zero even if I had seconds to spare. "GET BACK!", he would scream and then it was back into the head to start all over.

The torment went on for roughly thirty minutes before I began to choke and gag on the fumes which were rapidly taking the breath out of me, making me dizzy, and causing my nose and eyes to leak like a faucet. He continued screaming at me as I screamed back, "AYE SIR!", as loud as I possibly could.

Another thirty minutes and I was beginning to lose my bearing. It was one of those moments where the senior DI was needed to keep the kill hat DI in check, just so that he didn't take things too far and kill my ass by "mistake". It happened on Parris Island from time to time - where a DI would take things too far and shove a recruit into an industrial dryer or push a recruit off the third deck (it happened in 2016 long after my tenure on the Island). The senior DI would come in and scream, "STOPPPPPP!!", if he thought the recruit had enough or if the DI was taking things too far. The recruit would then stop, get to the position of attention, and scream, "STOP AYE, SIR GOOD MORNING, SIR!" and if the recruit screamed loud enough the senior DI would tell the recruit to go away back to their spot on line.

The problem was that on that morning, the senior DI was nowhere to be found. I was completely at SSGT Stubb's mercy and there appeared to be no end in sight to the torture I was being forced to endure. I again ran my way, thighs burning and lungs screaming, to the front hatch in a pushup position only for him to again scream, "ZERO! GET BACK!" I stood up and threw my cover onto the ground screaming, "FUCK!" That was gasoline onto the fire to SSGT Stubb (or any DI for that matter), and it prompted him to respond in a thundering voice, "GOOD, BITCH! LOSE YOUR BEARING TOO!"

I ran back into the head and began again, screaming as loud as possible the whole time. My throat was beginning to bleed by that point, and I could no longer see normally due to the fume vapors which were constantly soaking into my eyes. I made it halfway up the squad bay again when my arms gave out and I planted my face straight into the chemical solution. I quickly got to my knees and moaned, "OH GOD!!", as I attempted to wipe the burning liquid from my eyes. Hearing me call out to God in desperation caused SSGT Stubb to freeze and scream, "STOPPPP!", pausing only for a moment before he continued, "BITCH…THERE IS NO GOD! SCREAM IT!"

During that time in my life in recruit training, I was still holding onto a notion of God. I wanted to believe that he was with me during the trials I was facing and that his strength would guide me through to the other side, unscathed and heroic. I refused screaming back, "No, SIR!"

"BITTTCCCH!", he barked angrily, "SCREAM THERE IS NO GOD!" I knew that he would not stop until I did, so I obeyed the second time and screamed back to him, "SIR, THERE IS NO GOD!"

"BITCH! IM GOD! SCREAM IT!"

"THE DRILL INSTRUCTOR IS GOD!"

He screamed one last time, "BITCH, SCUZZ!", ordering me to continue with the torture to which I promptly did for another thirty minutes or so. No matter what he threw at me, I was NEVER going to quit!

At some point I remember some random DI who I did not recognize entered onto the squad bay which caught SSGT Stubb off guard. He told me to stop as he went to the front hatch to speak with the DI. I have no idea what they said, but looking back on it I assume the DI was most likely walking around outside and heard me screaming bloody murder from the barracks and decided to come up to see what was going on. I imagine that he warned SSGT Stubb about taking things too far and that he could be risking having an officer or some other high rank official stumble across the games we were "playing".

I assume this because after the DI left, SSGT Stubb came over to me and instructed me to stop what I was doing and immediately run along with him out of the squad bay. He told me not to scream and to be "tactical", meaning I was to make no noise at all. I'm guessing it was because of whatever the DI had told him otherwise I most likely would have been continuing that game for who knows how long. He then ran me out of the rifle range barracks and onto the range where he then proceeded to make me perform an incoming fire maneuver which consisted of me hitting the deck, getting up and running for three or four seconds, and then hitting the deck again (simulating incoming fire or artillery).

I did that maneuver down the whole 500 yards of rifle range until I got to the pits where the rest of my platoon was. SSGT Stubb then Disappeared, and I joined my platoon mates where I was greeted by one of the other recruits – the guide of the platoon - who wrinkled his nose when I approached. "Love the smell of bleach in the morning", he chuckled as I began pulling targets along with him. Once in the pits, the rest of the day went by normal… well normal for bootcamp.

RIFLE QUALIFICATION

During firing week, following afternoon chow, my platoon would take the range and shoot the targets for points and qualification. The first few days of firing week consisted of "pre-qualification" to see where we were shooting and to sight in the rifle as accurately as possible to maximize our scores. The goal was to get enough points to qualify as "rifle expert", but one only needed to get enough to score "marksman", which was the lowest score a recruit could obtain. Recruits who scored as a marksman were made fun of and were often called "pizza box" recruits since the marksman badge was square and resembled that of a pizza box – clever right?

The first few days of pre-qualification, I was shooting pretty much average. It did help that the DIs were not on the firing line and instead replaced, temporality, by Marine marksmen. The marksmen were much more relaxed and did not yell or scream at us like the DIs did and that put all the recruits at ease. Their only goal was to get us up to speed on the rifles, how to shoot them, how to group shots, and ultimately score as high as we could. The DIs would threaten us all at night about scoring high and they would mess with the recruits that scored the lowest that day – you know, just as incentive for the rest of us to make sure we shot as well as we could.

On the last two days at the range, we shot for actual qualification. I remember a photographer was out there taking shots of all the recruits for the graduation book. I finished shooting one of my sets and turned to reload when I noticed a camera lady pointing right at me. I remember having a brief thought and decided to smile. I figured if I smiled, they would most likely include the shot in the book and sure enough they did. I may have been the only one smiling in that whole book, but there I was, on my way to becoming a Marine and I knew I wanted that photo for later.

It was on the rifle range, late one evening in the squad bay, when a recruit by the name of Ackintosh broke down for no apparent reason and attempted suicide. I was not sure what happened, but I remember that evening when the DIs were playing various fuck-fuck games with us, he began to break down and then randomly lost his cool. The DIs had put him in a restrictive hold, sort of like a modified Full-Nelson (which we all later dubbed the Ackintosh hold), and it appeared to have a devastating effect on him. Sometime later, during a quiet moment when we were all cleaning our rifles, Ackintosh attempted suicide and that was it for him. Ackintosh was quickly whisked away by DI SSGT Fierro while DI SSGT Stubb made crash jokes at his expense.

Later that night I remember Akintosh was sitting alone on the quarterdeck surrounded by his gear and clad only in a pair of green skivvies and the green undershirt . His sneakers had the laces removed so that he could not hang himself with them. He sat there and cried inconsolably for whatever it was he was going through. He had been a good recruit up until that point and for whatever reason he snapped and was dropped from the platoon and the USMC for good.

The last day on the range I shot well enough for the sharpshooter badge and was in fact only about 10 points or so from shooting expert, but I didn't really care; I was just glad that I didn't fail the qualification or "Unq", as the DIs would say. That night I remember being on fire watch and writing a letter to my family to let them know how things were going. The whole platoon was silent, and an eeriness filled the air. Those rifle range barracks were old – at least 1960s – and I swear that I could feel the spirits from recruits and Marines past. As I got back into my rack on the bottom bunk after my fire watch had ended, I stared at the iron bars which lined the bottom of the top bunk and noticed something etched on the iron. I saw written in faded black magic marker – *HALFWAY THERE!*

DENTAL

We woke up earlier than usual on the day we left the rifle range in order to hump back to our original barracks. That morning was profound to me because it was one of the only times SSGT McHolland acknowledged my potential as a good future Marine.

The morning started as usual with head calls, chow, and morning clean up, however we also had to pack our large ruck sacks and stage them on line in front of us for inspection before the hump. My rack mate, a guy by the name Wilkins, had not finished packing any of his gear when he was pulled by one of the DIs to head outside to either get smoked in the pits or do some other task.

Once I finished packing my bag and staging it, I took it upon myself to pack my rack mate's bags and pack it properly with the proper weight distribution aimed at providing maximum comfort on a long hump. I then staged it on line next to mine and stood online at the position of attention. I was the first one done while everyone else was still hurrying to finish.

SSGT McHolland emerged from the DI hut and began screaming at us to get on line. As the recruits scurried to finish packing, SSGT McHolland walked by me and noticed that I had packed Wilkin's bag. SSGT McHolland himself had made Wilkins go outside earlier so he knew that he had not had time to pack his own bag. He stopped in his tracks and looked at me in disbelief. "Did you pack that bag, Verlice?" he barked.

"YES, SIR!"

His eyebrows raised and he let out a, "Hmmm." It was sort of like the sound one makes when they are inquisitive or pleasantly surprised at a task someone had performed. "Damn, Verlice", he said as I stood at the position of attention with my head and eyes to the front, "I figured you would have just said fuck him… okay." He took a short pause, raising his eyebrows once more, "carry on!"

It may not sound like much, but it was – it was to me. It was clear to me that he was pleased with what I had done and while he wouldn't admit it, his voice couldn't hide the truth. I had never seen a DI compliment a recruit, and I there I was having just received one (sort of) from the DI who hated me the most. As we walked outside in formation, I carried Wilkins's bag on my front, along with mine on my back, and couldn't have felt prouder of myself.

The hump back to our original barracks was more laid back

than the one we had initially made two weeks prior. The DIs had us in "tactical" mode which gave all our voice boxes a break from the incessant screaming. It was also at a much more normal pace than the previous hump and was somewhat enjoyable. Once we returned to the barracks, the next week was essentially uneventful with typical bootcamp activities such as marching, more chow, more "fuck-fuck" games, uniform fittings, and dental appointments.

By that point in bootcamp, we were beginning to get used to the flow of things. I started to realize that the DIs torment could only go on for so long before it would be time for chow again. Recruits learned to tell time by looking at the sun in the sky and eventually we got into the "chow-to-chow" living mode.

The recruits learned to live that way and eventually we learned that we wake up, get fucked up by the DIs and then go to chow; then we march for a while and then go back into the barracks for more "games" - maybe kick some classes, and then back to chow again for lunch. After noon-time chow, we would march some more, play more "games" and before long, it was back to the chow hall for evening chow. After that, more marching, more games, back into the barracks, shower, free time (one hour), and then hit the racks to do it all over again the next day. It became second nature by the end of second phase and truly signified that the transformation from recruit to Marine was almost complete.

In the beginning of the last week of second phase, several of the recruits including myself were ushered to the base dental clinic for some involuntary dental work. Unbeknownst to me, they had decided to remove all four of my wisdom teeth to which I would have preferred not happen, at least not while at bootcamp. I was whisked into a dental chair, given one shot of Novocain, which was nowhere near enough, and then had all four of my wisdom teeth yanked out of my face.

It was unpleasant to say the least and on top of that, the DIs would refuse to let us take any medication, not even Tylenol. I was marched back to the squad bay after the procedure and was literally dripping blood from my mouth. Try screaming with a swollen face and bloody holes in your mouth – like I said, no mercy given! I was on a liquid diet for the rest of the week and had to dig food out from the craters in my mouth for at least a month after that.

Shortly after the dental fiasco, perhaps only a day or two later, we were marched to an on base facility which was essentially a large warehouse. It was where they made us get into a big line while civilian women fitted us for our dress uniforms. I remember that day because it was the first time I had heard any music in almost two months. It made me feel somewhat human again to hear the lyrics, "*We fell in love in a hopeless place...*" echoing around the spacious warehouse. For a moment, I felt free again. Little did I know, things were going to ramp up in the next week once we entered the third and final phase of bootcamp. I was ready to get it over with.

2D PHASE III (WEEKS 8-12)

BWT

By the time we entered into phase III of bootcamp, the recruits and platoon as a whole were really beginning to come together. We marched more professionally, took the DIs punishments in stride, and even began looking the part. Our uniforms were now complete with name tapes, bloused boots, and the guide even carried the red guidon which displayed our platoon number, platoon 3080, which let everyone on the Island know that we were third phase recruits and almost ready to become Marines.

To begin phase III, the DIs started off by piling us all onto a white prison-type bus which then drove us out to the old, abandoned air strip (Paige Field) which was situated in the swamps of Parris Island. It was where we would eventually participate in the crucible in only a couple short weeks but was also where we did land navigation, patrolling exercises, and BWT. By that point in recruit training, we had just completed our final PFT (physical performance test) which included pull ups, sit ups, and the timed 3-mile run. The highest score one could achieve on the PFT was 300 (considered a 1st class PFT score).

In order to get a 1st class PFT score, a recruit had to do 20 pull-ups, 100 sit ups, and run 3 miles in under 18 minutes. I could easily get 20 pull ups and 100 sit ups but could never get my run that fast. I ran my final PFT run in about 23 minutes which was pretty average considering they gave recruits 27 minutes to pass. The running time was important because it factored in to how they broke us up for the BWT training.

BWT was only about two days and was essentially a mini crucible. BWT stood for basic warrior training, and it consisted of sleeping in the field, eating MRE's (meal ready to eat) for the first time, performing various firearm patrols, land maneuvers, and finally the BWT combat endurance course at the very end. The first day of BWT was not too bad and consisted of basic land navigation, basic patrolling techniques, eating MREs, partaking in various classes, and sleeping in these hooches which were essentially just small shacks with dirt floors. It really wasn't that bad; however, the last event of BWT, the combat endurance course, ended up being one of the, if not *the most*, difficult thing which I endured – physically - on Parris Island.

COMBAT ENDURANCE COURSE

I remember the night before the combat endurance course as all of the recruits were lying in the dirt inside one of the hooches, the chief drill instructor, SSGT Biggins, came in and addressed us in a very stern and intimidating manner. He came in an announced, "tomorrow is going to be one of the worst days you've had so far at bootcamp."

That really did a number on my mental psyche and caused my imagination to run wild with all types of gruesome scenarios about what the DIs would have us do. Shortly thereafter, I remember falling into a very short and unsatisfying sleep.

The one thing I learned at bootcamp (and the Marine Corps in general) was that they NEVER told us what was going on until we were literally about to do it. I almost guarantee during WWII, some of those Marines that stormed Guadalcanal and the other Islands in the Pacific, had no idea what they were about to do until that morning before they did it. That is how the Marine Corps worked in a nutshell, and like everything else with the Corps it's tradition that is passed down era after era.

The morning of the combat endurance course, the DIs broke us up into groups according to our run time on the final PFT. A third of the recruits with run times between 18 minutes to 20 minutes were in one group; another third of the recruits were put into a second group comprising of those that ran between 20 minutes and 24 minutes; and the rest of the recruits with the slowest run times were in a third group. It was set up that way so that the recruits would at least be able to keep up with each other during the combat endurance course.

I was, of course, put into the group with the recruits who ran the fastest and of which I had almost zero chance of keeping up with. I believe the move to put me in that group was most likely done as a result of SSGT McHolland's great disdain for me and furthermore, it aligned with his one and only mission: kill recruit Verlice. Also, guess what? SSGT McHolland was leading the group, so I was basically fucked right from the start.

The combat endurance course began with each group running in different directions for about a mile before getting to the starting part of the course. Once reaching the course, we all had to traverse these different obstacles such as crawling under muddy water through barbed wire, climbing various obstacles, monkey bars, casualty carries, and the course ended with another sprint to the end. When SSGT McHolland took off, we all followed suit attempting to keep up with his ridiculously fast pace. The guy was on a full dead-ass sprint, and everyone left me in the dust because again…they all ran faster than me. There was no way I could keep up with those speed demons as they all had an unfair advantage of faster run times.

I ran as fast as I could but just couldn't keep up. Luckily, SSGT McHolland, or anyone else for that matter, didn't notice that I was not with the group and eventually I was able to reach them just as they were beginning the course. I thank God that they didn't notice how far behind I was. I remember thinking that I could easily slip into the woods and escape the Island; that is how far back I was from the group. I was already half dead by the time I reached the course and things only got worse from there.

We low-crawled through 100 yards of mud and barbed wire and then had to climb a huge wall which required us to boost recruits in front of us – using teamwork – to get to the top. I remember I had slung my rifle around my back as other recruits had done as well and as I was climbing the wall, another recruit's rifle barrel smacked me in the mouth drawing some blood. I remember thinking how horrible the course was already starting out and how much I was just wanting to get it over with as quickly as possible.

We continued through the course and at some point, we began to switch off carrying recruits as if they were casualties. To carry a casualty required us to pick them up and haul them on our shoulders in a maneuver we called "fireman's carry". It sucked bad, especially if one was tasked with carrying a recruit that outweighed them by 40-50 pounds. At one point we got to a place where we had to take our rifles out and patrol as if we were being attacked by ambush. I remember a recruit we had nick named "puddles", (actual name was Perrier) for his uncanny habit of pissing himself all the time, began to vomit from pure exhaustion.

Eventually we finished the course, but SSGT McHolland was not satisfied, and he made us run the thing a second time. So, not only did we have to go through the bastard once, but we had to sprint the whole thing again, and if a recruit was not toasted by the end of the first run, they were by the end of the second. Looking back, I am surprised that I was even able to make it through twice given the speed everyone was running. At the end of the second run, almost half of the recruits were limping.

By the time it was over, the white buses came back and picked us back up, driving us back to the squad bay where we took a 30 second shower and then marched to chow. It had been one of the most physically exhausting days which I had experienced at boot camp thus far and gave me a good depiction of what the Crucible was going to be like. I fell asleep like a baby that night after lights out and had a vivid dream of arriving back home in my dress blues as everyone looked at me in awe – a hero's return!

REVENGE OF THE PONCHO!

Remember the poncho incident from the 1ˢᵗ day on the Island? If not, go back to the beginning and re-read to refresh the memory banks because that bastard caused me a lot of issues one morning shortly after BWT. That morning happened to be the morning we were going to march over to a building to get our graduation pictures taken - you know, the ones in the dress blues that everyone wants. On that morning, just like every other morning, the DIs got us up and we stood on line and proceeded with the morning as usual with head calls, clean up, and getting dressed for the day in our uniforms. It was the one and only morning that it happened to be raining the whole time I had been on Parris Island. When SSGT Stubb screamed for us to get our ponchos and bring them on line and hold them straight out and parallel to the deck, I immediately thought to myself, "*FUCK!*"

I had completely forgotten that I had never received a poncho and somehow it was missed or not noticed by the DI's when they had us mark all our gear on the first day. I had made it almost the entirety of bootcamp without needing the damn thing and now it was finally going to bite me in the ass. I quickly went back behind my rack to where my seabags were stowed and pretended like I was searching for it as SSGT Stubb quickly counted us all down to get back on line. I was hoping by some miracle that a poncho would magically appear in my sea bag – a gift from the Marine Corps Gods, perhaps; however, no such miracle occurred that morning. I knew that I was in for a painful lesson in gear accountability – one of the hallmarks of recruit training.

I returned back on line before SSGT Stubb got to "ZERO" and was the ONLY recruit not holding a fucking poncho in his hand! I was, again, the odd one out. There was nothing I could do; I couldn't simply explain to him what had happened and how I was never given a poncho and had ultimately been too intimidated by the DIs to ask for one on the very first day we met them. To the DIs, I had simply just lost the poncho, which was a capital sin of bootcamp. One NEVER loses gear! Losing gear was almost as bad as capital murder to a DI.

When SSGT Stubb saw that I did not have a poncho held out in my hands, he made a sharp beeline towards me and screamed in my face, blasting me with spit, "BITCH GET YOUR FUCKING PONCHO!" I screamed back, "THIS RECRUIT NEVER RECEIVED A PONCHO, SIR!" It was about as honest as I could be, given my set of circumstances.

He responded by punching me hard in the gut, and screaming again, "BITCH, FIND IT!" I screamed back, "AYE, SIR!" and then went back to my seabag to pretend to look for something which I knew wasn't there. SSGT Stubb did not give two fucks about my situation, and according to him I had better simply, "... SHIT OUT A PONCHO OR SO HELP YOU GOD...!"

I desperately attempted to look for it, dumping the contents of my seabags onto the deck while SSGT Stubb continued to scream in my ear, "FASTER FASTER FASTER!"

After several moments, he left and then out came SSGT McHolland. "Yep, here we go!", I thought to myself, feeling my heart pound heavily in my chest. I could only imagine what he was going to say or do to me, but I knew it wouldn't be good.

SSGT McHolland approached me and screamed for me to get on line in front of him at the position of attention which I promptly did. "Where is your fucking poncho, Verlice?"

"Sir, this recruit never received a poncho!"

He wasn't having any of it and screamed, "You're a fucking liar, Verlice; and when I go back to check the gear list that every recruit signed on the first day and I see your mark next to a poncho, I'm going to punch you in the mouth!" He then left and I knew that due to the "fog of war" during the first day we met our DIs, I had put my mark next to that piece of gear out of complete confusion. He basically had me dead to rights and even though I seriously never did receive a poncho, it was going to look as though I had just lied to him. I was scared shitless and knew I was going to have to deal with his wrath once he returned.

Several moments passed and sure as shit, SSGT McHolland

emerged from the DI hut holding in his hand the piece of paper which I am assuming was the gear list that had my mark made next to a poncho, making it appear as though I had received one. "YEA RIGHT, VERLICE!", screamed SSGT McHolland as he zipped towards me. "WHAT THE FUCK IS THIS?"; and with those final words, SSGT McHolland reared back and punched me directly in the mouth. Blood poured out onto the deck and onto the paper which he held in his hand. "GOOD!", he barked, "NOW I HAVE IT MARKED IN YOUR BLOOD!" I swallowed the blood that pooled up in my mouth as I screamed back, "AYE, SIR!"

He continued screaming at me angrily while I stood at the position of attention, "AND WHEN WE DROP YOU FOR BEING AN INTEGRITY VIOLATOR, YOU WILL KNOW WHY!" All I could do was scream back, "AYE, SIR!" before he angrily departed back into the DI hut.

I remember some of the recruits around me were showing concern given the fact that my mouth was bleeding. It was unusual for a DI to hit a recruit in the face as it was a dead giveaway of abuse – things they weren't supposed to be doing in the first place! SSGT McHolland had clearly gone too far and I think he even knew it himself. One of the recruits close to me, Bibliogi, gave me a tissue for the blood coming out of my mouth. I was furious and really felt like hitting SSGT McHolland back, but I remember Bibliogi calming me down and saying, "It's not worth it, Verlice."

After several moments, SSGT McHolland came back out with a poncho in hand and threw it at me. He made me put it on like Little Red Riding Hood and then proceeded to make me low crawl up and down the whole squad bay while he basically ridiculed me for being a piece of shit. "If I was you... I would beat this bitch every night, everyday!", screamed SSGT McHolland as I low crawled up and down the squad bay trying desperately to hold the bloody piece of tissue paper on my lip.

After several moments of ridicule, he ordered me to get back on line and then he again disappeared into the DI hut as all the recruits and myself stood on line at the position of attention. I still had the bloody piece of tissue paper on my lip but eventually threw it behind me on my rack. I began to worry about what he said earlier - about dropping me as an integrity violator. That was something that I did not want to have happen and was something that they definitely did if the need arose. If I had been dropped then, I'm not sure if I would have been able to start the process over. I may never have become a Marine.

Eventually the Senior DI, came from the DI hut and barked out my name. I ran to the front quarter deck where he simply examined me without saying anything. Looking back, it was clear he was checking out my lip to make sure that it wasn't noticeable; however, he didn't say it outright and instead made it look like he was just angrily observing me. SSGT McHolland had actually been lucky because when he punched me, my tooth went through the inside of my lip and other than being a bit swollen on the outside, it could not be detected otherwise. If he had split my outer lip open, I always wonder what they would have done. I would have had to lie and say, "Hey, look... I slipped in the rain room..." or something if an officer had ever questioned me about it.

After "checking" my lip, SSGT Fierro proceeded to smoke my ass on the quarterdeck while screaming at me about how great of a Marine SSGT McHolland was. He was going on and on about his exploits in Fallujah, Iraq, and his other accolades. I think he was worried that I might tell someone – which I would not have done – and wanted to remind me about the measure of a Marine such as SSGT McHolland.

After about a half hour of being smoked on the quarter deck, SSGT Fierro ordered me back to my position on line, and then we proceeded to march outside and eventually to the building to get our graduation pictures taken. I never heard about the incident again and they did not drop me. I think they considered the situation a draw because if they had attempted to drop me, I would have simply told the officers that I had never been given a poncho and had been punched in the mouth and *I KNOW* for a fact that was not something the DIs wanted to be promulgated. So, I got my picture taken with a slightly swollen lip and that was the end of that.

BEFORE THE CRUCIBLE

Before platoon 3080 could complete bootcamp, there were a few loose ends that needed tightening up. First, we took part in team week which meant all the recruits were split up into teams and sent around the island to do various work details and other boring things like that – nothing too crazy. Then we prepared for final drill by practicing the marching and rifle maneuvers repeatedly in order to compete for 1st place, which we did not get. We also performed the final CFT (combat fitness test), final USMC written knowledge test, and some other basic housekeeping duties before setting off on the final test of recruit training: the Crucible.

I also remember that it was around that time that SSGT Stubb blasted me with so much spit that it should have constituted assault. It happened during evening prayer while I was huddling with my prayer group right before lights out. I had innocently reached behind my head to scratch an itch when SSGT Stubb, quick as a flash, grabbed me by the collar and screamed, "STOP MOVING YOUR GODDAMN BODY!"

He had screamed so loudly that a large quantity of spit flew from his mouth and ended up dangling off the end of my nose and as a reflex, I promptly wiped it off. Seeing that, he instantly screamed at me again, blasting me with more spit which collected on my forehead! Knowing I could not wipe it off again, I had no choice but to simply let it slowly dribble down my nose, over my lip, and down my chin while everyone in my prayer group tried their hardest not to laugh. It was all about discipline; a recruit never moved unless they're told to do so.

THE CRUCIBLE

On the morning of the crucible, the DIs woke us up early at roughly 2 am or so. It was a little bit different than usual because the DIs, I noticed, were not wearing their usual smokey-the-bear campaign cover but instead had on their regular garrison cap covers like the rest of us. That made them look less intimidating which was part of their MO during the crucible – observing rather than ordering. The point of the crucible was to put us all to the test and for them observe us to see if we were able to apply everything that we had learned throughout the course of bootcamp.

It was the final test before we became Marines, and it was

going to be a tough one and I knew it…everyone knew it. The crucible lasted 54 hours and consisted of many physical and mental exercises meant to test our level of knowledge and Marine Corps physical fitness. Over the course of those 54 hours, with only 10 or so hours of sleep and only 3 MREs, we would be led through various stations designed to test us in every way possible. It was meant to show the DIs that we as recruits were finally ready to bear the title United States Marine.

Over the course of the Crucible, we would be hiking and running a total of over 50 miles, tested on weapons maneuvers and tactics, and pushed to our absolute limits with a final 12-mile hike back to the Iwo Jima monument to receive our EGA (eagle, globe, and anchor) emblems. Receiving our EGAs would be a ceremonious moment and would signify our transformation from recruit into Marine. Everyone, including myself, was excited but nervous to begin the crucible. I think we were all ready to get the whole thing over with by that point.

We started out by hiking all the way out to the abandoned air strip, Page Field, where we had done BWT several weeks prior. I don't remember the hike out there being too taxing as we were all pretty much energized and ready to start. In only three short days we would be Marines, but I knew a lot could happen in between during the crucible, and I was apprehensive about how it would play out. One thing I knew for sure was that nothing was going to stand in my way of becoming a United State Marine and even if I had to crawl my way back to the Iwo Jima monument, I was going to finish what I had started.

When we got out to Page Field, the DIs again broke us up

into groups like how they did during BWT, and I thank God that I was put into a group which was being led by my senior DI. That group of 15 recruits would be the group I would spend the entire 54 hours of the Crucible with, and it honestly could not have been any better for me. If I had been put in SSGT McHolland or Stubb's group, I'm not too sure it would have been possible to get through. That may be why SSGT Fierro opted not to put me in either one of those groups – because even he knew they would have destroyed me. I mean, either way I would have made it...but seriously: thank God himself I was not with SSGT McHolland.

COOPER COLOR CODE: BLACK

Once the groups were together and broke apart, SSGT Fierro looked at me and said, "Verlice?"

"Yes, Sir?"

"My goal is to get you to Cooper Color Code Black during these next three days."

The Cooper Color Code chart was something we learned in our USMC classes, and it basically stated the levels of pain, disorientation, and weakness that a Marine could feel during combat or some other demanding event. Cooper Color Code black was the worst level and basically meant that a Marine was almost dead – physically and mentally broken. So, that was his goal: to get me to Cooper Color Code black, and he'd be dammed if he wasn't going to make that happen. The man ended up being good on his word.

The first day of the Crucible was not too bad from what I

remember as we were all full of energy, sleep, enough food to keep us going, and because the different stations were not that difficult to overcome. We ran here, ran there, carried each other on our backs, simulated various famous USMC firefights, practiced ambush skills, patrolled, did land navigation, and performed various teamwork and team building exercises all while the DIs stood back and observed, involving themselves very minimally. They really wanted to see how we were coming together as a group of soon-to-be Marines. Each station of the Crucible was named after a different Marine who earned either the Navy Cross or Medal of Honor in some famous USMC battle such as the battle of Hue City, Fallujah, and many others.

It wasn't until later that night, around midnight, when we were told we could only sleep for 3 hours that things began to get more difficult. The DIs were not telling us when to eat, so it was up to us to make the MREs they provided us last the whole time, and some recruits chowed theirs down before the first day was out. Everyone was starting to get hungry, tired, and sorer by the hour as the Crucible was raging into the second day.

I remember getting very little sleep that first night and dreaming about what I would do when I got to go home for the 10-day boot leave I would receive after completing bootcamp. I had a girlfriend named Gabby whom I had not spoken to since I left, and I was anxious to see if she had waited for me or if "Jodie" had already gotten to her. She was on my mind a lot during those three months of bootcamp even though it was one of those teenage romance-type relationships which was destined to fail anyway.

We woke at zero dark thirty and began the day with more

running and more teamwork stations similar to the day before. It was on that second day when the totality of the Crucible began to hit me, and I began creeping closer to that Cooper's Color Code black. We slowly began to run out of food and everyone, including myself, was beginning to become disoriented and dehydrated. Halfway through the second day we had to do an event where we ran up a hill with our rifles in full combat gear and shoot at targets in a combat-engaging exercise. We then had to run through a trail in the woods which was physically taxing due to not only the weather but also the 60 pounds of gear we had on our bodies. It was after that event that I truly got to Cooper's Color Code black.

DI SSGT Fierro ran us to a station which was called the "John Quick Trail". The station was named after a US Marine who had won the Medal of Honor during the battle of Cuzco, Cuba on June 14th, 1898. The whole point of the station was to be led through a trail while the DI would simulate various attacks on the patrol. The point of the exercise was to test our patrolling techniques which we had learned throughout bootcamp. We had to simulate taking contact from the left, right, front, rear, and react accordingly. It was supposed to be a laid-back patrol but quickly turned ridiculous after the first "contact" was made.

SSGT Fierro did not like how we reacted to one of the simulated "attacks" and began to make us run through the trail at full speed. Now keep in mind, I was already half dead from the whole day's events as well as the prior days, and this was just further taxing my weakened body. Every time we took "contact" and didn't react quick enough, SSGT Fierro made the punishments worse and worse until I was literally crawling on the ground with all my gear and another recruit on my back, straddled like a backpack. I was crawling an inch at a time, heart pounding, sweat pouring, and my body at the breaking point.

I was beginning to lose my composure and noticed for some reason that I was pretty much the only recruit that was being punished so hard by the DI. The other recruits were still upright running while I was crawling with a whole human on my back (to simulate a casualty). The recruit on my back, I don't remember who it was, kept whispering, "I'm sorry, Verlice", because he knew how close to the brink I was. I can remember screaming back to him, "JUST SHUT UP!" as I continued to gruel my way through the dirt and mud.

We finally finished by sprinting back to the air strip where SSGT Fierro then lined us up and asked if any of us were close to passing out. One of the recruits said, "Sir, check on Verlice!"

I was literally wobbling back and forth and while I didn't think I was going to pass out, I did feel as though I would collapse from pure exhaustion. DI SSGT Fierro looked at me, saw my condition, and nodded, an impish grin forming on his face, "Good!".

Well, he had achieved it; he had gotten me to Cooper's Color Code black.

ALMOST DONE!

After the John Quick Trail fiasco, we took a short break to hydrate and eat what was left of our rapidly dwindling MREs. I felt a lot better and was able to continue with no medical attention.

After a short break, we ran the combat endurance course again; however, we walked through it "tactically" and as a result it was not as bad as it had been during BWT. I don't think I would have had the energy to run through that course again at full speed multiple times. That night we were up until about midnight doing the final event, the night movement course, which entailed traversing under barbed wire, over obstacles, and performing sprints with ammo cans weighing 40-50 pounds. Gun fire sprayed over our heads and flares lit up the night sky as we low-crawled our way through the course.

After the night movement course, we marched back to the hooches and were told to sleep for a few hours so that we could wake up early to begin the final march back to the Iwo Jima monument. I was in so much pain by that point that I was seriously dreading the hike back. We only slept 3 hours or so before we were woken up for the final hike.

FINAL HIKE BACK

When we woke up, I would like to say that I was feeling energized and ready to go, but to be frank I was feeling like a bag of shit. I was wondering if I was going to make it back to the monument in the condition I was in. Before we took off, I thought it would be a good idea to eat my last MRE for a final burst of energy; however, it ended up being a mistake on my part. My last MRE was the dreaded Vegetable Cheese Omelet, which was dubbed "the vomelette" for its uncanny ability to make a recruit or Marine sick as fuck. They had been discontinued and I'm assuming the one I had was a little devil that slipped through the cracks somehow. I remember eating the "vomelette", which literally looked like a pile of vomit, and also scarfing down the dried granola and "powdered" milk that came with it. Keep in mind, I ate that fucker cold!

How is that for hard core, huh?

It wasn't long after eating the vomelette that I began to feel sick to my stomach. *Great!* It was all I needed as I was staring down the barrel of a 12-mile hike in full gear. I already knew that the final hike back was going to be one for the books.

As we started walking, the regret from eating the MRE became more and more prevalent as the nausea in my gut began to intensify. It wasn't quite to the point where I was going to puke, but it was a thought in my mind as a possible likelihood. The hike back was not a leisure stroll in the park either; they were not taking it easy on us and for the first 6 miles we were practically running the damn thing primarily due to the dreaded "slinky effect".

During the final hike, recruits were supposed to be close enough to the recruit in front of them so that if they reach out their hand, they could touch their back. I was feeling like such shit that I was more than a few arms distance from the recruit in front of me and because of the sick feeling in my gut, I was struggling hard to keep running and touching their back. Luckily, I never puked but if I was going to, I would have had to literally puke as I ran to keep up. There was absolutely no stopping for any reason. You want to shit? You're doing it in your pants. You need to puke? You're doing it as you run or walk. No mercy was ever given even in those last moments of being a recruit. Thankfully, the feeling eventually went away, and I was then able to keep up normally.

FRUIT AND A GATORADE

About halfway through the final hike, the DIs told us to stop and allowed us to sit down on our packs for a few moments. It was a welcome break as I was still slightly recovering from the bad MRE I had eaten before stepping off. I remember the sun was beginning to come up over the horizon, and God was it a glorious morning! The sun was burning a brilliant red and orange and as it rose a bit higher in the sky, so did my spirits. I knew that the ordeal was almost over and that in just a couple hours I would be calling myself a United States Marine. The DIs made several recruits hand out an apple, banana, and a Gatorade to all the impoverished recruits and we all wolfed it down in record time. I swear I had never eaten anything so delicious as that banana was that morning. The fruit and Gatorade saved me, and I was instantly pumped full of so much energy and motivation that a freight train wouldn't have been able to stop me from finishing the final hike.

THE FINAL STRETCH

After we finished our short break, we began the hike again and I felt so energized that I damn near could have run the rest of the way to the Iwo Jima monument. As we continued along the road, we eventually began to hike into the main part of the depot where I could then see all the buildings and squad bays and at that point, I knew that there was only a few more miles left to go.

The hike had been mostly silent up to that point but once we were about two miles away from the monument, SSGT Fierro began to shout out the very first cadence I had heard since being on the Island.

"OH, WHOA OH WHOA!", he shouted loudly as we all echoed it back with enthusiasm. "OH, LORDY DON'T YOU KNOW!"

I was feeling alive and full of energy as he continued. "WHEN I WAKE UP IN THE MORNING…. THERE'S A DRILL INSTRUCTOR THERE!" He went on with the cadence, lifting all our spirits as we echoed back his words proudly.

"A DRILL INSTRUCTOR… A DRILL INSTRUCTOR… OH WHY WONT YOU GO HOME…AND LEAVE ME ALONE!" Each and every recruit screamed back his cadence as loudly as we could, hoping everyone on the Island would know that we were about to become the nation's newest Marines.

I remember a recruit who was marching behind me, Williams, was struggling badly at the tail end of the hike. He kept saying he was going to pass out, that his feet were killing him, and that he could hardly walk anymore. He was beginning to fall out as he limped as fast as he could to keep up. "Grab my bag and I'll help you the rest of the way there", I said, feeling the spirit of Chesty Puller himself coursing through my blood.

He grabbed my pack, and I pulled him along with me; I even passed him my canteen so that he could drink some of my water. Williams held my pack the rest of the way to the Iwo Jima monument. That is what it was all about; that was what all the training was for! By that point, I knew that I had earned the title of **United States Marine.**

CHAPTER 3

A RECRUIT DIES A MARINE IS BORN

After another mile, we finally reached our destination: the Iwo Jima monument. The monument was situated in front of Peatross Parade deck where our platoon had performed final drill just a week before and where in only one more week we would finally graduate. There was not one soul in platoon 3080 that did not want to get off that God forsaken island.

The DIs formed us up into several large formations so that each platoon of Mike Company was facing the monument. Once we got in formation, we dropped our packs, slung our rifles, and stood at parade rest while the battalion commander came out – don't remember his name – and gave a little speech to congratulate us on finishing the crucible and becoming the nation's newest United States Marines.

I don't remember all of what was said as I was overtaken not only by exhaustion and hunger but also emotions. There was not a dry eye in the area, and it was the first time I cried while on Parris Island. Becoming a Marine was the proudest thing I had ever done in my life. I had been a directionless, unmotivated underachiever and now I finally felt like I had a purpose – a mission – in my life. I felt like I could move mountains that morning standing in front of the Iwo Jima monument, dirty and exhausted... but alive!

SO HELP ME, GOD

At the end of the speech, the company commander had us all raise our right hand and repeat our oath as United States Marines to defend the United States of America.

I (Charles Franklin Verlice) do solemnly swear that I will support and defend the Constitution of the United States against all enemies, foreign and domestic; that I will bear true faith and allegiance to the same; and that I will obey the orders of the President of the United States and the orders of the officers appointed over me, according to regulations and the Uniform Code of Military Justice. So help me, God.

After the oath, we stood back at parade rest and were then ready to receive our EGAs from the very DIs who had been tormenting us for the last 12 weeks. Each DI went down a different row of recruits and it was Staff Sergeant McHolland himself, the DI who hated me the most and wanted nothing more than for me to have dropped out, who handed me my EGA. When he got to me, I held out my hand, looked him straight in the eyes as tears rolled down my face, and shook his hand. "Congratulations, Verlice", he said in the most human voice I had heard from him the whole time during bootcamp. It was over; I had made it! I remember thinking in my head at that moment,

Wow! I actually did it!

The only thing that was on my mind after receiving my EGA was graduating and getting off Parris Island. I was also looking forward to seeing my parents who were coming for family day even though, admittedly, I was not that home sick.

After the EGA ceremony things were immediately different and it was like weight had been lifted off my shoulders. We were no longer treated like recruits; we were going to be treated like low-level Marines. In my case, I was a private so that really wasn't much higher on the ranking scale, but I was still a Marine, so that had to count for something in the eyes of the Marine Corps, right? The DIs, for the very first time since I had been on Parris Island, allowed us to walk back to the barracks on our own out of formation. It was the first time I had walked somewhere by myself as we were always in formation up to that point and it felt very weird to be away from the DIs. All the screaming, "fuck fuck" games, and everything else was over, and we even began calling our Drill Instructors by their rank instead of "Sir".

On my way back to the barracks, I helped a Marine who was hobbling from blisters and leg cramps get to where they needed to go. Once I got into the barracks, the first thing I did was take a piss in the head without a DI screaming behind my back. It was glorious! I realized how much I took the little things in life for granted before joining the Marine Corps. Even something as insignificant as taking a piss or shit without someone glaring at me or screaming at made me feel as though I had won the lottery. After I did a few things in the squad bay, I went back outside and walked to the chow hall where the ceremonious "Warriors Breakfast" was being held. The warrior's breakfast was something that every new Marine was awarded after completing the crucible and becoming a United States Marine.

WARRIORS BREAKFAST

I remember not going too crazy with the food during the Warriors Breakfast like some of the other new Marines did. Some of them literally ate until they puked which I thought was both stupid and pointless. Everyone was allowed to talk amongst themselves, sit normally while eating, and go up for seconds or thirds. I got a big glass of chocolate milk for the first time since the DIs would never let us drink it during recruit training. All the DIs sat at the table and ate along with the new Marines; some of them even asked normal questions and spoke to us like human beings.

A few high-ranking officers also sat down and asked us various questions about our experience during training as well as our thoughts on earning the title. I remember I was sitting next to my buddy, Bibilogi, and we were talking about how cool infantry school was going to be when SSGT McHolland came behind us, put his hands on both of our shoulders and said aloud, "these are going to be two of the finest 03's the Corps has ever seen!". He was referring to our MOS's as 0300 infantry Marines. That was a huge compliment to me and coming from SSGT McHolland, the DI who tried hardest to destroy me, it meant even more and is something I still keep with me to this day.

MARINE WEEK

The last week of bootcamp was called Marine week and it basically consisted of returning gear, attending some classes on what the next steps of our career in the USMC was going to be, and of course receiving our orders to our next duty station. For me, it was Camp Geiger to attend ITB (infantry training battalion), since my job in the Corps was going to be infantry. Also, the first Sunday after becoming Marines, we were afforded a four-hour liberty on base where we could go to the MCX (Marine Corps Exchange), buy candy, food, and other pointless things, and of course make a phone call home. I was excited to phone home as I had not been able to since the very first night I had arrived on Parris Island.

Before we went on liberty that Sunday, we were told by the DI's that we were to go in groups or at least have a buddy. I ended up going with my friend, Bibilogi. It was truly liberating to be walking around the base free of a DI and with a bit of money to buy a few things. It was in fact the first time in my life that I had ever had any real money to my name and I felt like a rich man walking around that MCX where the first thing I bought was a huge bag of peanut M&Ms which I had been craving the whole time I was on the Island. I also had to buy a watch – a G-shock – which the DIs ordered us all to have by the time we returned back to the barracks. It was nice to have a watch and finally be able to tell the time and not have to guess by looking at the position of the sun in the sky.

Next, we went to a building where there were wired phones on the wall. Nobody was allowed to use or have a cellphone on Parris Island and in fact I never even brought one with me, so the only way to make a phone call home was to use the landlines provided at base facilities. The line for the phones was so long that we spent most of the four-hour liberty waiting to get on one so we could make a quick one-minute-long phone call home.

When I dialed the house phone that my mother had installed before I left for bootcamp, my father answered and we chit chatted for a few moments about how I had done, and he of course reaffirmed to me that they were going to be there next week for family day to attend my graduation. And that was that; we went back to the barracks and had to surrender any candy or other food items we had bought. For me, it was my giant bag of M&Ms which later ended up in SSGT McHolland's hands that night as he was on duty.

FAMILY DAY

Family day occurred the day before graduation and began as they marched us over to a large hangar where inside contained bleachers for people to sit. They formed us up right outside the doors to the hangar and had us stand in formation for about 20 minutes before they let us march in. I remember I was in the last row and my DIs were standing right behind me also waiting for the orders to march in. I remember reaching behind my neck to scratch an itch and was quickly grabbed by SSGT Stubb who had been standing directly behind me.

He yanked me by the collar and scolded in my ear, "Do you know what I would do to you right now if you were still a recruit?" He said that because a recruit was NEVER allowed to just scratch or move while in formation and even though I was a Marine, it didn't matter. The guy was still crawling up my ass and treating me like a recruit.

"Yes, Staff Seargent!", I responded back.From the corner of my eye, I saw SSGT McHolland shake his head to SSGT Stubb essentially conveying the message, *Yea, we're not doing that anymore.* It was like even he knew SSGT Stubb could take things too far.

We were finally ordered to march inside the hangar and set up in front of all the families who were sitting in the bleachers. They did a speech, introduced our DIs to the families, and then broke us free for family day. I remember it took me several minutes to find my parents, but eventually I did, and it was quite nice. We hugged and they asked me various questions about what the experience was like and all I could do was enthrall them with various stories about what I had endured and what I had experienced at the hands of the DIs. I do remember being very paranoid that my DIs were going to tear through the crowd, find me, and scream at me in front of my parents. I knew that some of them were itching for it, but thankfully they never did.

My sister, who had accompanied my parents, let me see my old phone which I had given to her before I left for bootcamp. It was one of those old cell phones that slid up to reveal a keyboard which I had purchased from Wal Mart after my high school graduation. I got on Facebook, posted some cheesy status about my accomplishments, and of course messaged Gabby.

I remember we walked around looking at the various things to see around the Island as well as going into various stores and the museum they had there. I remember feeling so bad for my parents because they were so broke that they could barely afford the trip to see me and furthermore, could not afford to buy my little brother, who was about five at the time, a toy from one of the stores. I bought it for him with the scant amount of money I had earned during bootcamp.

After family day was over, we gathered in our platoons near the parade deck and practiced the final graduation march and ceremony activities about 50 different times. The DIs almost seemed to be putting on one of those "dog and pony shows" for the families who were still watching us from a distance by making us run all over the place every time we would mess up. It was kind of annoying, but I didn't care too much since I was about to graduate the next day.

That night, I held my final fire watch on Parris Island, and I remember thinking about everything that had happened thus far during my Marine Corps experience. I was excited and anxious about leaving the island but was also nervous about what the rest of my time in the Corps was going to look like. I fell asleep that night feeling as though I could walk on water.

GRADUATION

That morning, we woke up super early and marched over to a chow hall which we had never been to. There was a new platoon of recruits waiting to enter and they must have been in first phase because the DIs for that platoon were smoking the hell out of them. They had one of the new recruits in front of a tree screaming at it over and over for about 20 minutes. "SCREAM UNTIL THE TREE SCREAMS BACK!", they demanded without mercy. I felt *so* glad that I didn't have to deal with that kind of nonsense ever again.

After morning chow, we went back to our barracks and dressed in our service uniforms which on that day were the green Alpha's. We then staged our gear outside and marched over to the parade deck for the ceremony. The sun was already out by the time we got to the parade deck and all the families were staged in the bleachers waiting to watch the graduation. There was a marching band playing various tunes and everyone's spirits were soaring.

We began the ceremony which lasted some 20-30 minutes and consisted of the new Marines marching around the parade deck, various Marines being given awards such as the highest shooter, highest PFT, and guide recognition, and final march around the Peatross Parade Deck. They then had the guides of each platoon retire the guidons, and we made one last march in front of the crowd. The last thing they did was give us our final orders: dismissing the platoon. SSGT Fierro did an about face and announced to the platoon,

"PLATOON 3080, DISMISSED!"

Each of us, in unison, screamed, "AYE, STAFF SEARGENT!", and then did an about face before screaming, "OOH RAH!"

That was it. Bootcamp was officially over.

The band began to play as all the families ran up to meet their Marines. I remember at that moment an odd and ominous feeling washed over me; it was like a feeling of dread and terror. I felt different; I felt changed. I couldn't explain it, but somehow, I felt something bad was lurking on the horizon. I should have been happy and filled with pride but instead it was like my gut was telling me something awful was going to happen. I felt it, but I just didn't know what it was – I couldn't quantify it. The universe just said,

No, you wait. It's coming, Verlice. Everything you asked for is coming right to you!

My family finally met me and snapped a few photos of me right there on the parade deck. I remember being so anxious to leave that I was rushing them off the parade deck and to the lot where my father had parked the rental car. We made our way to the car where I had previously stowed all my gear – my seabags and uniforms – and finally drove off the Island. I had done it and was finally on my way home. I then felt like I could relax!

CHAPTER 4

HOME, IF ONLY FOR A MOMENT

The first stop we made outside of the Island was the McDonalds where three months prior I had smoked my last cigarette. I remember being resolute that I wasn't going to smoke cigarettes ever again since we were not allowed to during bootcamp, however I quickly broke that rule and lit one up right there in the parking lot. Afterwards, I went into the bathroom at McDonalds and took off my Alpha uniform, stowed it neatly in my bags, and put on my civilian clothes which I had been wearing the very first night I arrived on the island. I was back in my lucky red button up, jeans, and New Balance sneakers – now called go-fasters. I remember feeling so lightweight since I was accustomed to the heavier materials of the USMC cammies.

Back outside and thinking I was so cool with my dog tags

hanging out of my half buttoned-up shirt, I was approached by another new Marine fresh off the Island who was also in civilian clothes. He whispered in my ear, "Tuck those dog tags in…they're watching us out here, you know?" I remember how crazy that sounded, and as soon as I realized that I was tucking my dog tags in too I thought, *Wow, they really did brain wash me!*

When I finally got home, nothing felt the same. It was like I was home, but I wasn't. It was bizarre. I remember checking my weight in the bathroom to find that I had gained some 15 pounds during bootcamp. It felt weird to watch TV again since I hadn't seen one in three months. Even listening to music was weird.

The next morning, I tried to keep to my bootcamp routine by waking up early and going on a run, working out, and eating "chow". It was in fact the only time I adhered to my bootcamp routine while I was on boot leave as I quickly fell back into my nasty civilian habits. Using my old phone, I got into contact with a couple of people I had been friends with before leaving for bootcamp and told them all about my experience trying desperately to impress them with all of my feats only to find that none of them really seemed to care.

Soon I got into contact with my girlfriend, Gabby, and I told her everything about bootcamp. She expressed interest in wanting to continue our relationship; however, it was on a very loose basis and even though I kept in on again off again contact with her for the next year or so, the relationship never grew to more than that.

I bought a few things while I was home including a new

cell phone. It was 2012, so it was one of those rudimentary "smart phones" which one could get at Walmart, but it was better than anything I had before the Corps. I remember being so gung-ho and brainwashed that I was literally marching around in the Walmart like I was back on the Island. That was something that I would *never have* done before since I cared way too much about what people thought of me but that was how strong the Marine Corps' effect was on me. It was like they had injected me with some phony tough guy courage.

At some point during the week, my family took me to the mall which was about an hour away from the house I spent most of my adolescent years in, and there I took pride in walking clad in my issued USMC tracksuit. I wanted everyone who saw me to think I was a hero who had just got back from war but in truth, I was just a boot and had done nothing more than complete bootcamp.

My mother bought me a zippo lighter which had my name "Verlice", and "USMC" engraved into it. I kept that lighter with me the whole time I was in the Marine Corps and still have it to this day. I remember my father also bought me a little Marine Corps button which I ended up leaving on the seat of his car by mistake. Eventually, years later, he returned it to me, and I also have that to this day as well.

A day or two later, my parents drove me to Eglin Air Force Base where I was so proud to show them my military ID and show off the fact that I was a Marine. I got a haircut on the base and made sure to have them do it really "high and tight" because I wanted to exuberate Marine Corps spirit to all those around me. I found myself wanting to show everyone around me that I was a Marine but not realizing that there was more to it than just bootcamp. I had not thought about anything during boot leave: not ITB, not the fleet, not deployment or anything in between. I was truly flying blind on the journey I did not realize (or grasp) that I was on my way - via one way ticket - to a world of pain that I never could have fathomed.

On the last night home, I spent my time packing all my uniforms and gear up for the flight to North Carolina where I was to report to Camp Geiger for infantry training. I was nervous to fly since it had been years since I had been on an airplane, but I suppose I was more nervous about going to infantry school since I really had no idea what to expect.

My mother drove me to Pensacola International Airport on the morning I was flying out and it couldn't have been any somber of a drive. I said my goodbyes to her and went through the gate to wait for my plane to start boarding. I remember having this bad feeling; a feeling that I may never see her again which I know sounds dramatic, but it was how little I knew about the Marines. I knew that I would not get leave for a while and as far as I knew, I would be sent straight over to Iraq or Afghanistan once I was done with infantry training. As I sat waiting for my plane, I texted Gabby and looked at random things on my phone. Eventually they called for my plane to start boarding and as I stood up, seabags in hand, I remember thinking, *Alright. Let's do this.*

CHAPTER 5

SCHOOL OF INFANTRY (SOI)

CAMP GEIGER

As I flew my way to North Carolina, I kept wondering what infantry training was going to be like. Surely it couldn't be as bad as bootcamp, right? I mean, it's not like I was going to be dealing with DIs anymore. I was a member of the club - I was a Marine - and I expected to be treated as such. I knew the training would most likely be tough, but I just hoped there wouldn't be the abuse of boot camp. I really didn't think I could take any more of that.

Halfway through the flight, I began to develop a bad migraine. It was one of those migraines that wouldn't quit; even Tylenol didn't ease the pain. After about eight hours or so of flights, layovers, and driving in a cab, I eventually found myself on Camp Geiger where they dropped me off at a holding platoon with tons of other Marines who I had been to boot camp with. I remember meeting a couple of my friends from my platoon and "smoking and joking" with them at the designated smoke pits (gazebos where people smoked).

I remember feeling great, besides the migraine, and was happy to be able to go to the chow hall to eat food – I didn't say good food, just… food. I remember thinking, *I could get used to this.* Well, unfortunately, that was the one and only time I ever ate at a chow hall on Camp Geiger as infantry training did not allow such luxuries. I would find that out in just a few short hours.

That night, we all slept in one of the holding platoon squad bays which was a bit different than the types of squad bays they had on Parris Island. They were still open with bunkbed racks lined up, but they had these huge wall lockers which sort of separated each bunk so that there was at least a modicum of privacy. I will admit, I wasn't impressed with the accommodations and found them to be quite poor – only barely livable if I'm being honest. I mean, what was I expecting right? I fell asleep with my head pounding and full of thoughts.

What is this going to be like?

ITB

So, let's go over a bit of housekeeping just so that what I will be going into in the coming chapters won't be too confusing. ITB, short for Infantry Training Battalion as it was called back then, is where any Marine with the MOS (Military Occupational Specialty) designation '0300 – pronounced OH -

300" is sent to for specialty training. ITB was an eight-week course which got all infantry Marines combat ready before sending them to a deployable unit in the Fleet. The Fleet is what all Marines call the main pipeline of the USMC. It was the real Corps outside of bootcamp and any subsequent training a Marine must undergo, and it was said to be way better being a Marine in the Fleet than a Marine in the training pipeline. The Fleet is where a Marine would be stationed and the unit they would eventually deploy overseas with.

Any Marine who was not in a contract 0300 infantry MOS position went instead to MCT (Marine Combat Training) which was like a watered down four-week course where they ate a couple MREs, shot a few guns, and then shipped off to their next school whether its air winger school, communications, intelligence or any other non-infantry school. MCT Marines went through this baby course to keep aligned to the Marine's saying, Every Marine Is a Rifleman! Which is not really true because only an 0311 was considered a true rifleman. Any Marine who was not a grunt was affectionately referred to as a POG (Person Other than Grunt). The ITB Marines and MCT Marines never mix and outside of that very first day, I never saw any of them again.

Early that morning, some Sergeants woke us all up and marched us over to some big building where they began to section us off one by one. They split all the Marines up: if a Marine was MCT they went to one side of the room and the ITB Marines went to the other side. The whole process took about three hours and eventually we found ourselves just waiting for our combat instructors to come and get us and take us over to our permanent squad bays to begin the training.

After a couple hours, a Marine Sergeant came in from one of the side doors and said, "Hey, Marines! Where are my MCT Marines at?" A couple of the Marines on the MCT side said,

"Here, Sergeant"

"Well come on, guys! Let's get going. We're going to hit the chow hall really quick and then I'll show you the barracks and we'll get ready to get underway here!" He seemed to be in good spirits and was not giving off any DI vibes. You know, no rageful screaming, spitting, or "physicality" whatsoever. The MCT Marines got up and filed out of the building and then it was just us – just the ITB Marines left.

I was a bit more relaxed because at least they weren't screaming at anyone like they did back on Parris Island. I was hungry and felt like I could go for some of that shitty chow hall food.

What will I get? Maybe a salad or some chicken… who knows?

Several minutes went by, and then more minutes went by and before I knew it an hour had passed. I remember thinking, *damn, why are they taking so long?*

After a few more moments, the side door burst open and in popped a Sergeant – a black guy whose name I cannot remember. He stuck his head in and screamed, "GET THE FUCK UP AND GET THE FUCK OUTSIDE RIGHT NOW, PACKS ON YOUR BACKS IN FORMATION ON YOUR FACE – PUSH UP POSITION…NOW…NOW…NOW!"

The Sergeant's angry and demanding demeanor really took me off guard as I had been lulled into a false sense of security from the MCT training Sergeant. I remember thinking, *are you fucking kidding me right now?*

We all ran as fast as we could outside where, to my dismay,

114

it had begun raining - hard! We all got on our faces in a push up position with our heavy-ass seabags on our backs while the Sergeant screamed at us, telling us about how we weren't MCT Marines - we were infantry Marines - and that they were going to make damn sure we were put into as much pain as possible during the training.

He eventually made us get up and take our packs off and hold them above our heads and then squat down. He then ordered us to start duck walking here and duck walking there for about 20 minutes while we all got soaked and wet. I was really beginning to get envious of those MCT Marines and was wishing I had picked a different MOS - one that was a bit easier on the knees.

After a few more "fuck-fuck" games to which I was so familiar with by that point, he ordered us to run at a dead sprint. We had all our gear on our backs and were sprinting, about a mile, to our squad bay. I remember stepping in a pothole and rolling the fuck out of my ankle. I rolled it so badly that my ankle touched the deck. I thought I had broken it and had a brief moment of panic thinking my USMC journey could be over right then and there.

Luckily, even though it was painful and swollen, I did not break my ankle. An incident like that would have changed the trajectory of my USMC experience for sure. As I swallowed down the pain which was rising from my rolled ankle, I managed to limp-sprint the rest of the way.

Once at our destination, we were ordered to duck walk some more and then finally told to sit down on our packs while we were introduced to all the other combat instructor Sergeants. I remember getting irritated and thinking, *okay, starting to get hungry here! When the FUCK are we going to the chow hall?*

My question was answered after about thirty minutes when I saw a couple of the ITB Marines being told to bring a pallet of something over to all the other Marines who were seated in front of the ITB Sergeants. When they got closer, there was no mistaking it; I saw the white boxes…they were bringing in those damn bag of nasties which we had eaten in boot camp. You know, the dry bread subs and generic chips…yea, those fucking things!

OH, YOU HAVE GOT TO BE KIDDING ME!

No part of me wanted to eat that nasty shit, but unfortunately it was all we were going to be getting that day, so I had no choice but to force it down. The instructors made sure to inform us again that we were not MCT Marines and would in fact never get to go into the chow hall and that it was not going to be "summer camp".

After finishing those nasty box chows, the instructors passed around sheets of paper which was essentially a list of the 0300 jobs we could choose from. They told us to write the one we wanted the most at the top and so on down to the one we wanted the least at the bottom. The options were 0311 rifleman, 0331 machine gunner, 0341 mortarman, and 0351 assault man. Those were the only options given to us.

I remember SSGT McHolland told us he was an 0311, and I had remembered all the stories he told about his exploits in Fallujah, Iraq and the combat he had been in. I remembered how decorated he was and how he talked about the 0311 rifleman being the tip of the spear head. I wanted to be the one going street to street, room to room, kicking in doors and taking names. I put 0311 at the top of my list. Because there were only a limited number of infantry jobs outside of the 0311 designation, most of the Marines ended up being selected as an 0311 and if it was put at the top of a Marine's list, it was the job they automatically ended up getting.

So, I was selected to be an 0311 rifleman and the combat instructors put myself and all the other selected 0311's in three different platoons while any Marine who was selected to be either a machine gunner, mortarman, or assault man was put into their own platoon called "weapons platoon". Weapons platoon was OFP (own fucking program) meaning they followed a different training regimen than all the 0311's and we never mixed with them much during ITB.

After we were put into our platoons – I was in 3rd platoon, known eloquently in the Marine Corps as "turd platoon" for anything third being shitty in the USMC – the instructors had us clean the barracks spotless. The instructors pretty much left us alone during that time and were not up our asses like the Drill Instructors on Parris Island had been. We were not treated like recruits either; we were allowed to talk, use our cell phones, when not training or busy, and could make head calls whenever we wanted. We were still not allowed to leave the barracks without permission or run wild, but the freedoms were significantly more than bootcamp allowed for.

The combat instructors were still very demanding of us; however, the purpose of their mission was to get us all combat ready and not necessarily just haze us pointlessly like Drill Instructors did. I've always maintained that bootcamp was harder mentally while ITB was harder physically just due to the constant running, hiking, and field ops. As ITB began, I was beginning to realize that I was in for some of the most difficult physical training that I had experienced up to that point in my life.

TRAINING

The next day we began our training which involved going into the field for the first time. The first time in the field lasted for a week and for the whole week it rained. Essentially every week, besides the last two weeks, was spent out in the field from Monday through Friday when we would then return to the barracks for the weekend. The last two weeks of ITB was the final infantry test and it was 2 weeks long and tested us on everything we had learned in infantry school. While out in the field we dug fighting holes and slept in them, patrolled objective points endlessly, fired weapons of every kind at all times of the day and night, and also learned other infantry tactics including ambushing and long-range patrols. We ate only MREs and basically lived as if we were on a deployment.

The first day we went to the field, we threw grenades which I remember being somewhat worried about just because of the odd chance that it might blow up in my hand and kill me. Obviously, that didn't happen, and it ended up being a bit anti-climactic and relatively uneventful. I just threw it and heard it explode, seeing the big burnt hole it left in the dirt. That night, we slept in a bivouac area in the middle of the woods inside holes we dug prior. I remember it was March, so it was still cold – cold and rainy. I remember not being able to stay warm no matter what I did. Those were the moments when we all really started building that Marine

Corps camaraderie since we essentially needed each other for body heat. The whole "oh, that's gay!" thing goes out of the window quickly when its freezing cold, wet, and you're hungry and tired.

The very next morning in the field, the combat instructors woke us up by firing rifles over our heads and throwing grenades somewhere in the distance to startle us all. When we were "ambushed", we all had to grab our rifles, which were usually slung around us even in slumber, and attempt to form a 360-degree defensive position. Being brand new to the infantry life, of course we failed and then had to be "taught a lesson" which consisted of us being ran in full gear from here to there. That was usually the go to punishment in infantry school: you mess up, you run.

Later that day, I remember one of my combat instructors saw a Marine slacking off with his rifle barrel sort of jammed in the dirt. I remember he zipped over to him, grabbed him, and body slammed him onto the ground, screaming at him to never let his rifle touch the deck. It was the last time he, or any of us for that matter, ever made that mistake. That was essentially how the combat instructors treated us – rough and harsh but with purpose.

The rest of that week consisted of learning how to patrol properly and what to do in the case of an ambush. We also dug 6-foot-deep fighting holes and slept in them. There were two to a hole and one person had to always be on guard all night long in case of a simulated "ambush". We switched off with each other and at around 3 in the morning, the instructors, dressed in camo suits, ambushed us from the woods by firing shit tons of blanks at our positions.

They kicked a lot of classes with us while we were in the field. The classes were akin to the same type of classes we would take in bootcamp where we learned everything about the Marine Corps; however, the classes in infantry school were more in depth and geared towards rifleman tactics, rifle knowledge, and all pertinent knowledge that an 0311 would need to know to accurately perform their job.

When we were in the field, we did nonstop patrols where one squad would sleep, one would be on guard, and another would be on ambush looking for various targets. We switched off and a Marine could expect to sleep maybe 3 hours or so a night. It was very exhausting, very physically demanding, and very monotonous. At the end of the first week, I was ready to get back to the barracks and get weekend liberty, or "Libo" as we would call it. They let us have "Libo" every weekend, unless we did something to lose it, and that meant that we could take one of the ridiculously priced taxis into Jacksonville and go to the mall or any number of other boring places.

The instructors would disappear on the weekends, and we would basically have the squad bay to ourselves. Of course, there always had to be Marines on "duty" to watch over the barracks and we would draw straws or something to decide who was the one who would be stuck with it. Duty sucked because we couldn't leave the barracks, had to remain in uniform, and basically just sat at a card table and did nothing until another Marine came to relieve us.

I didn't draw duty that weekend and I remember getting my phone back later that day, which the instructors made us keep in a large Ziplock bag during the week. I remember going out into town with a buddy of mine who I'll call Holdin. Holdin had been in my boot camp platoon and was in fact the scribe of my boot camp platoon. I had never really talked to him or even really liked him that much during boot camp; however, I realized in infantry school that we had a lot in common. He reminded me a bit of my cousin, and we just sort of clicked and got along well together.

Often, we would go out into town and frequent the Jacksonville Mall, which was a common stomping ground for ITB Marines and fleet Marines alike, and just wander around aimlessly. Sometimes we would eat somewhere or watch a movie at one of the theaters there and basically just waste time. Anything to get off Camp Geiger was good enough for me and it didn't matter if I was sitting around doing nothing or watching a movie. I remember Holdin being somewhat of a shy and awkward guy with a quirky sense of humor. He was from somewhere either in Mississippi or Alabama and had quite a southern accent. He was also a very soft-spoken guy and often made awkward comments which I found funny. I can remember him talking with his girlfriend on the phone and would have to refrain from laughing at their weird conversations. I would be able to hear her flirting with him or being sexually provocative and his responses were so nerdy and lame that it was hard not to chuckle.

It was also nice to be able to sleep in on the weekends and because of all the physical rigors of ITB, it was very welcome. ITB is really where I began to smoke heavily as most ITB Marines and fleet Marines alike are heavy smokers or dippers (chewing tobacco). I would often chain smoke up to a pack or a pack and a half a day depending on how busy we were. At the end of the day on Sunday, the combat instructors would come back into the squad bays and demand that we put our cell phones back into the big zip lock bag so that we would not have them during the week. It always sucked doing that because it was like they were taking away our lifelines – our only connection to the civilian world.

Even though I was an 0311, we still had to shoot every weapon including the 240s. The 240 was a big rapid-fire machine gun and was heavy as hell to carry. We had to carry these things everywhere we went while in ITB too. Often, we would run here and there, and everywhere we went our gear and rifles went with us. Sometimes, the instructors would make us stage all the weapons and then take us into a building to show us some videos of people being shot and other "motivational" war-type footage. I'm assuming they did that to get us into the mind-frame of a killer. They wanted us to *want* to kill and when it eventually came time to do our jobs, they wanted to make sure that we would perform without hesitation.

SGT GEORGE

It's funny because for some reason I cannot remember any one of my combat instructors' names…except one: SGT George. SGT George was a stand-up guy - very well put together, and just an all-around stellar Marine. I remember he was on the verge of being accepted to OCS (officer candidate school) to train to become an officer. SGT George was a true Marine and is the type of Marine I expected to meet during my time in the Marine Corps. Unfortunately, SGT George was one of the only decent Marines who was ever in charge of me during my time in the Corps.

During the second week of infantry training, on a day when

it was raining, wet, and cold, several other Marines and I got into a bit of trouble for breaking away from the rest of the platoon during some down time and sleeping under a tree to keep out of the rain. Apparently, it was something they did not want us to do, and once they caught us, they punished us quite severely by making us grab heavy boxes of ammo and chow and run them all over the place until we were half dead. They also informed us that they were taking our liberty for that coming weekend and the news of that almost broke me as I lived for that liberty. I remember SGT George approaching me and asking me if I was okay because I was having a little bit of a moment where I was doubting myself and my decision to not only be an 0311 but to join the Marines in general.

I remember he pulled me aside and talked to me one on one – man to man, telling me a very personal story about when he was a young Marine and how he had doubts and troubles too. He talked to me like the father I had always wanted. Part of the reason I joined the Marines was to find direction in my life – to find that father figure that I didn't really have in my own father. To me, SGT George was that father figure, and I would have liked to have had more Sgt Georges in my life during my USMC experience. After speaking with Sgt George, I was motivated and never felt that doubt while in infantry training again. It was like he uplifted me and gave me the juice I needed to get through the rest of the training.

It was during that same week that we learned how to do land navigation, often dubbed "land nav" for short. They taught us how to read a compass, shoot an "azimuth" (don't ask me because I still don't know what the fuck it is), and read a topographical map. Once we learned these basics, we were sort of put into the middle of very deep and dark woods and had to use our map reading skills to find different boxes which were hidden in various locations. The activity was done in squads of around 7 Marines and took most of the day to complete.

I can remember finding those boxes in the woods – there were maybe 15 of them in total – and every time we found a box we noticed that each one had a penis drawn on it. Every subsequent box we found had a more detailed penis drawing on it and when we finally found the last box, I swear, you would have thought Leonardo DaVinci himself drew that cock. It was hilarious and was something we found out quite quickly as infantry Marines: everything will at some point have a cock drawn on it.

During the first couple weeks of ITB, we learned about and shot almost every type of weapon from the M16-A4 to the 240, grenade launchers, AT-4 rocket launchers, and everything in between. We learned extensively about booby traps and IEDs (improvised explosive devices), which were the biggest killer and maimer of Marines in Afghanistan and Iraq. I remember IEDs being the one thing I was afraid of the most. I wasn't as afraid at the prospect of dying but more at the prospect of being maimed and coming back home with no arms or legs.

The combat instructors did not play around with us and if they caught wind of any of the ITB Marines messing around, being lazy, or trying to skate (get out of doing something), then they would take it upon themselves to make sure that nobody would try them again. I remember one evening while we were in the field on one of the weapons ranges, during some downtime while we were supposed to be eating our MREs, a few of the Marines decided it would be a good idea to have a competition.

The point of the competition was to take several Marines who did not dip and have them put shit tons of chewing tobacco in their mouth. The point, I guess, was to see who would vomit first and the one who *did* vomit first was considered the loser of that round. I remember not wanting to watch it or take part in the "game", so I sat around the corner with a few other Marines eating MREs. After a few rounds of the game, our combat instructors saw what they were doing and grabbed the last two Marines who were "competing" and then forced them to chug canteen after canteen of water before making them run around in circles for hours. It caused those two Marines to vomit copiously for the rest of the night and also put a stop to the little "games" some of the Marines were having. Some things like that I never understood or got the point of.

Probably around the third week or so of ITB, the instructors rallied us all together and told us that they were going to hold a "smoker's challenge". The objective of the smoker's challenge was to gather several of the Marines who were the heaviest smokers and grill them, quiz them, and fuck with them pointlessly and the winner got to bond with one of the instructors by smoking cigarettes together. It was sort of a comradery moment, and we all took it seriously because everyone wanted to be on the combat instructor's good side. I happened to be one of the Marines, along with six or seven others, who were selected to participate.

Those of us selected to participate were all directed over to a barren area and then ordered, one by one, to stand in front of all the combat instructors while they essentially just hazed us and ridiculed us for their own amusement. When it was my turn, I found myself standing in front of all the instructors feeling about as nervous as possible due to the obvious disparity between student and instructor. They essentially just made fun of me like crazy and asked me the weirdest questions – the type of questions which I had no answers to. They asked me things like, "if you had to choose to fuck one of us, which one of us *would* you fuck?" and "which one of us do you want to kill?" and other questions designed simply to amuse them and fuck with me. They laughed and mocked me each time I responded to one of their questions.

One of the questions they asked me was, "Which one of us is the best father figure?" and I remembered the conversation Sgt George and I had had a couple weeks prior. When I announced, "Sgt George, because he cares like a father." I could see in his eyes that it touched him – it meant something to him. It was just an unspoken moment between two people, and I'll never forget the look on his face – a mix of shock and touched. All the other instructors were sort of mocking me and laughing with each other as I murmured my response, but Sgt George didn't. To me it almost looked like he was going to cry...or that he was at the very least choked up.

As I said before, Sgt George, to me, was a true Marine. He was what I thought all Marines should have been like and exuberated all the characteristics which I believed Marines were supposed to have. Earlier when I had my moment and was seriously contemplating quitting, instead of berating me, beating or assaulting me, or mocking me, Sgt George showed compassion for me as a fellow Marine who needed help, and he took it upon himself to fill the role of motivator. It was directly because of Sgt George that I made it through infantry school and if not for him, I wouldn't have this story to tell. Even though my time in the Marines turned tragic, I am so thankful to Sgt George that he was the Marine he was and wherever he is now, I hope he remembers that moment.

GUNS, GUNS, AND MORE

So again, ITB was full of shooting guns, taking guns apart, humping the guns on long hikes, sleeping with the guns, eating with the guns, shitting with the guns... you get the idea. Those things went with us everywhere we went without exception! During ITB, we would have these conditioning marches where we would be in full combat gear with our rifles and packs, Kevlar and flak jackets and bags. The humps in ITB were much more intense and were significantly longer than the humps we had in bootcamp.

The instructors were not easy on us during these humps either. For example, on one of the hikes – a 13-mile hump – we did it at such a fast pace that almost every Marine was broken off physically. Once we got back to the squad bay late that night, we were told to sleep for a few hours. I remember it was one of the few times that my feet were torn from severe blisters and I could barely walk. The blisters, I remember, were so deep that the blood had soaked through the thick canvas material of my combat boot.

The next morning at around 6 am, the instructors woke us up and informed us that we were to be outside for a 7-mile combat run in boots and utes (short version for "boots and utilities" meaning we wore boots, Cammie bottoms, and green skivvy shirt) along with our rifles. I remember thinking, *there is no way that I am going to make a 7-mile run in combat boots with these blisters on my feet.* I tried to tell one of my SGTs that I had bad blisters, and he essentially told me to go fuck myself. There was no getting out of anything in ITB! It ended up being one of the hardest runs I had done up to that point in my life and the fact that I even made it through was a miracle. I limped most of the way, but hell… I did it.

HYNZE

I made several good buddies while at ITB but one guy I got closer with than any of the others. This guy's last name was Hynze and he slept in the rack directly behind me in the squad bay. Early on in ITB, we quickly found that we had a lot in common and as a result we became almost like brothers. He was almost a mirror image of myself in terms of personality, likes, dislikes, and character. I always considered myself to be a weird dude and very seldom have I met someone like me – someone I can really relate to.

We talked about all types of things and had the same interest in women, music, food, and even had the same devious con-artist type way of thinking. He was a bit older than me, 22 or so, and was from Utica NY. I remember he was a reservist, so after ITB he was going home to check into his reserve unit, meaning I would most likely never see him again.

Hynze was a sneaky character and would boast to me about how he was stealing various pieces of gear to take home with him as trophies. I never knew if it was true or if he was just trying to seem cool, but I believed him. I remember one weekend, before lights out on a Sunday, he handed me a bottle of Gatorade and told me to take a sip. I opened the bottle and took a sip only to be hit with a strong burning sensation in my throat. "What the fuck is that!" I coughed. He laughed like a lunatic and remarked, "It's moonshine!" The bastard brought moonshine into the barracks! An act like that would have gotten him in deep shit with the instructors had they found out and only God knows what they would have done to him. That was Hynze, though – one sneaky SOB.

THE WEIRDEST CHARACTERS

One thing I noticed quickly in ITB was how weird everyone and everything was. The military in general is a melting pot of people; people from all over the country mixing in one tight-knit community. People from the north, the south, the east, and west and everywhere in between. There were people from all types of cities and towns - big and small and some that came from shacks and even dudes who had no home. There were people with all types of stories, backgrounds, families, and education. The Marines is a smaller unit - there are less people in the ranks, and the infantry, being smaller than the Marine Corps as a whole, seemed to attract the weirdest of the weird.

The Marine Corps was not like anything I had experienced "on the streets" back home and during high school. Things like cliques, groups, gossip, and everything in between didn't exist there in the way it did back home and whatever a Marine had been "back on the block" did not apply to the Corps. I saw nerdy guys be the cool guys and cool guys be the wimps. I saw the weirdest guys at the top of the food chain and ex-starting quarterbacks in fetal positions sucking their thumbs and crying for their mothers. It really was just a total mind-fuck and bizarre space to be in because it flipped everything I thought I knew about social structures upside down. It almost seemed like the infantry was a haven for those weirdo types. The Corps was a place where repressed psychos and bullies could torment others and where those that had been bullied could avenge themselves.

Don't get me wrong, there were a lot of good guys too, but the vast majority were weird and off putting – like misfits. I always said that there were three types of people who joined the USMC infantry: Moto people, hurt people, and crusaders.

The moto people were exactly what one would expect: people who were all gung-ho and "OOH RAH", and "GET SOME" types. You know, the types you'd see in the gym lifting 500-pound weights and grunting at the top of their lungs while their buddies snort protein powder off the other's abs (I actually saw this happen).

The hurt people were just lost. They thought they could find themselves in the Corps and that the Corps would solve all their problems. Sometimes they were bullies or bullied people; sometimes they were just lonely people looking for a spot to fit in - looking for a club to be involved in.

The crusaders just wanted to make a difference; they wanted to see war and have that experience - stories to tell their grandkids one day. I always used to think I was a crusader. I wanted to believe that I was, but now I know that I was a hurt person. I was lost and thought I'd find the answers to all my problems in the Marines. I was looking for motivation, direction, and purpose.

There were tons of interesting people around me too, like a kid named Burr. Burr never spoke a word and would literally just sit and chew tobacco all the time. That was all he did when there was some down time – in the field or in the barracks. He would just sit there and flatten out the lids to the dip cans and add them to his huge stack. He would just hold his spit bottle and stare into space. Never even spoke a word to the guy nor did I hear him speak a word to anyone else.

Another guy, Vanhooten, used to run around the barracks naked and just headbang to some of the weirdest music I had ever heard. Dubstep was super popular around that time, and even though I had never heard of it before ITB, that was the type of music he would be head banging to. The guy would make the weirdest faces and just go into a full body convulsion whenever Dubstep music would play. Strange guy, Vanhooten.

Another Marine, last name Kyle, was the nerdiest guy I had known in the Corps. He looked like the kid from American Idol – the one they called "chicken little" back in 2006 or so. Although he was pretty beefy and relativity built, he was nerd through and through. Everyone would tease him by screaming his name in the middle of the night, "KYLE...DUM DUM DUM...KYLE!" Believe it or not, he was popular there in the Corps and fit in with everyone else perfectly. The kid who probably never got laid in high school or even had a girlfriend was literally a popular Marine in ITB.

The list just goes on and on. There were some people I hated, some people I liked, and some that I was indifferent to. It was in ITB where I first met a guy who ended up being a real close friend to me later in the fleet. His name was Castillio. He was a Costa Rican guy from New Jersey and was the shortest guy I knew while in the Marines. He was probably barely five feet tall but was built with a good amount of muscle. I had not really talked to him much until the last couple weeks of ITB, but it was clear that he was funny, and that we had a lot in common.

There was also a guy who I never spoke a word to in ITB but who would end up being my barracks mate in the fleet. His name was Drummer, and he was essentially a douche bag. I did not like him at all in ITB, but we ended up getting pretty tight in the fleet. That's how the Marine Corps worked: one moment you hate someone, then next they are your closest buddy.

It was just a strange place to be.

TERMINOLOGY

Another thing I found out during my time in the USMC was that even the language of the Marine Corps was different from anything I had experienced on the "streets" or during high school. The Marines had a term for almost everything. We called sneakers – go fasters; we called flashlights – moonbeams; we called the reflective belt – a glow strap; we called a pen an inkstick. They referred to us as Bitches, Pussies, Faggots, Queers, Weird-o's, Retards, Shit Bags, and everything in between.

If we looked disheveled, smelled bad, or our uniforms were not up to the USMC standards, they would say, "You look [or] smell, like a bag of smashed assholes!"

A Marine who was caught falling asleep during instruction, letting their head fall down before jerking it back up, was said to be, "bobbing for cock". Everything was always a threat too. For example, if we moved slow, didn't do what the instructors wanted, or just did our own thing that somehow did not align with USMC standards, then Uncle Sam would inevitably find a way to "get his" before the day was through. And rest assured...he always did!

Anything that happened to a Marine before they joined the Corps was said to have occurred "back on the block", and this saying was also used to refer to any Marine who did not uphold the strict USMC ethics and standards. A higher ranking Marine may say, "So, I guess Verlice just wants to do whatever he wants to do, right? I guess we think we are Billy Bad Ass back on the block, right?"

If we decided to be stupid and not clasp our Kevlar helmet under our chins, letting the straps dangle like a Marine storming Iwo Jima, then some eagle-eyed higher up would scream, "We're not fucking John Wayne here, Marine! Strap that Kevlar...NOW!"

All our girl friends who were still "back on the block" would inevitably end up cheating on us with "Jodie" – some nameless homeboy who roamed around looking for a lonely Marine's girlfriend or wife to fuck while they are away on deployment or training.

The instructors also used the saying, "good to go" for damn

near everything. It was not used in the same way that one would say it back on the block. For example, your boss might say, "Hey, Jim? Are you good to go for the meeting today?"

In the USMC, the saying was used anytime they wanted us to acknowledge something or do something in the affirmative. For example, a Marine may say, "Okay, Gents. Today we are going to assault this area, good to go?", to which we would all just sound off with a grunting response, "ERRR!"

"Then we will move over here and get chow, good to go? And then we will form up over here for this objective, good to go?"

"ERRR!"

It was like saying, "Do you understand?" It was just strange.

STRESS

ITB was all about putting us under stressful conditions. Whether it was doing a hump, a run, taking one of the many riflemen written tests, or many of the field exercises, the combat instructors were intent on making sure that everything we did was under the most stressful conditions possible. For example, one morning, they instructed us that we were going to run a CFT which is similar to the PFT only it starts with a half mile run, then proceeds on through some obstacles, where we had to fireman carry another Marine, and then finally ending with some other exercises such as pushups and "grenade" throwing. In some ways, the CFT was more physically demanding than the PFT even though there was technically less running involved.

On the morning that we were going to perform the CFT,

the instructors brought us out something they called "A-Rats" or A-rations which was essentially a hot chow. We didn't get A-rats often, but it was usually welcome when we did even though quite frankly, I thought they were disgusting. It came out in these large trough-like vehicles and consisted of powdered eggs, potato triangles, and some type of mystery meat that had a funky after taste. It was the type of food that "back on the block" I would never touch but had no choice but to in ITB since it was either that or nothing.

We all ate these A-rats and then immediately ran the CFT afterwards. I remember everyone was feeling horrible – sick and queasy. I can't be sure, but I swear they put something in the food that morning to make us all feel sick so that they could simulate extremely stressful conditions during the CFT. I remember barely being able to get through it because of how sick I was feeling. That was something they did often, whether running us until half dead on a full stomach before a test or "simulating" combat conditions before some physical exercise or evaluation, and it was one of the more difficult aspects of the training at ITB.

They would often make us perform an exercise called "stress shoots" while we were on our week-long field operations. A stress shoot consisted of simulating an "ambush" type scenario and then making us run for several miles while carrying heavy equipment or fireman carrying another Marine. They would then make us "assault" the objective by having us run at a dead sprint up a large berm where once at the top we would have a certain amount of time to hit all the targets with a certain degree of accuracy.

Every time they would make us do a stress shoot, I would be

half dead when I got to the top of the berm but somehow, I was always a perfect shot and never had a problem hitting my targets. I remember I would be struggling to breathe and in so much pain that it was the only thing I could think of. That is what they wanted – for the pain we felt to be worse than the prospect of engaging with the enemy. Marines are expected to fight until the last breath and never give up or quit, no matter what the circumstances.

Eventually we took our final "rifleman" test which was a written scantron-type test all about the fundamentals of being a rifleman, knowledge of infantry tactics, and all that good stuff. The test really wasn't that hard, but if one had ADHD like me and just didn't care to pay attention, it was still confusing. We were expected to know everything from how fast a 5.56 round travels, muzzle velocity, the max range for the M16-A4 and tons of infantry knowledge to which I don't remember most of today. After that test and another weekend liberty, we prepared for our final field exercise at ITB – the IFEX.

The final field operation was going to be two weeks and was going to test us to our breaking point as infantry Marines. It was essentially the Crucible on steroids. It even mirrored the Crucible by ending with a final hump; however, that one made the Crucible hike look like a cake walk. The final hike was over 20 miles in full combat gear after two weeks in the field subsiding off only a couple MREs. It was really going to challenge us as future infantry Marines.

IFEX

The IFEX or Infantry Exercise, as it stood for, was the

culminating event of ITB and the final event before graduating infantry school and being assigned our units. It begun with us being transported to an isolated area and then consisted of two straight weeks of field exercise, shooting, patrolling, night ambushes, and everything in between. Living off MREs, dealing with dwindling water supply, and really learning to lean on the Marine to the left and right of us. I would say that was where I really began to feel like I was a true Marine.

Most of it is a blur to me at this point, but I do remember being extremely tired and hungry and becoming somewhat dehydrated at some point towards the end. We were dirty, sore, and close to the breaking point. The event ended with the culminating 20-mile hump, which really ended up being a bit further as all the humps usually were.

We got ready for the hike at the end of the IFEX and took off at about 7pm or so. The hump took us all night to finish and there were several Marines that could not make it. One of the Marines, I can't remember his name, was in my ITB platoon and was actually washed out of ITB because he just could not complete the hump. I remember Hynze was behind me for about half of the hump and was struggling badly towards the latter half of it. He ended up falling back a bit, but he did finish along with the rest of us.

I was pretty good until the very last couple of miles when I then started to feel the pain and fatigue of everything. One of the worst things about the humps in the USMC was something I mentioned earlier called the "slinky-effect". The "slinky-effect" occurred because Marines at the front of the formation set the pace way too fast with their long strides and a Marine near the back of the formation, such as me, would have to run to keep up with everyone. Often, I found myself running the majority of the hump and if you think about running over 20 miles with full combat gear after two weeks of field training, you can understand how difficult that would be.

On the last mile, I remember my flak jacket was choking me and I could not breathe; my legs were so wobbly that I felt like a stiff breeze would topple me over. I was beginning to fall out and really didn't know if I was going to make it; however, one of my combat instructors approached me and motivated me to keep going by instructing me to loosen the flak around my neck which then allowed me to breathe easier and as a result, I was able to complete the hump and the IFEX, effectively completing ITB. It was one of the hardest humps I think I ever did my whole time in the USMC.

GRADUATION

After the IFEX, we were allowed to sleep until about noon since we didn't finish the hump until about 6 in the morning. There were still a few days left of ITB which was basically going to consist of getting our orders, returning gear, and then graduation. I remember the next day, Hyzne and I got into a stupid fight about something which I cannot even remember. He was also upset because he had wanted to be active-duty and was unable to for one reason or another, so for whatever reason he was taking that out on me. I remember we almost started fist fighting but ultimately did not, however I didn't speak to Hynze again until the last day.

During those last couple days, the instructors gave us our

orders to see what unit we would be sent to. All the 0311's – which meant me as well– were going to be stationed at a unit based right out of Camp Lejune, North Carolina. I was very upset about that because the unit was right across the "river" and meant I would be stuck in shitty North Carolina. Everyone in the Corps knew that Camp Lejune was one of the worst bases (outside of 29 Palms) to be stationed at and everyone wanted to be one of the "lucky ones" and get stationed in Hawaii. All the weapons Marines in my ITB training battalion were sent to Camp Pendelton in California to a different unit, much to my dismay. I was quite jealous of the weapons Marines and wished that I had chosen to be a mortarman or machine gunner, but by that point there was nothing I could do. I had made my bed as a rifleman and that was that.

Once I had my orders, the only thing left was to graduate ITB and go to my unit. I was apprehensive because the combat instructors told us story after story of how new "boots" in a unit got hazed. Hazing, while not allowed in the USMC – and was a zero tolerance – happened all the time and even the higher ups knew. Again, like in bootcamp, as long as an officer or high rank doesn't "see" it happen, then it *didn't happen*. They told us horror stories of boots being put in wall lockers and shoved down flights of stairs, new Marines being beaten, and everything in between. Everyone including myself understood that hazing was another rite of passage as an infantry Marine and I knew that I was a target to get hazed due to my age and my youthful high schooler look.

Over the next day or so, we returned all our gear which we had used during ITB and packed all the gear issued to us from bootcamp to take to the fleet. I remember the instructors also made us shave our heads before the graduation. I was irritated by that because I had started letting my hair grow back a bit, within regulations still, and didn't want to buzz it all off as I felt like it gave me some of my individuality back. They did it on purpose; they wanted to ensure that we were going to go to the fleet looking like a boot. Most guys in the infantry, especially some of the older salts, grew their hair out as long as they could within regulations, and it was the hallmark of a boot to have a shaved head. In the infantry, a boot had to earn the right to grow their hair out and this was, again, another rite of passage in the USMC infantry. Essentially, the combat instructors wanted to make sure we knew our place once entering the fleet.

So, with freshly shaved heads, we woke up early on graduation day and performed a final run. It was about three miles, and we sang the weirdest cadences. One of them went, "I USED TO FUCK THREES AND FOURS…NOW IM FUCKING TENS AND WHORES…" to which we all let out a chuckle.

Another one they sang was this super motivational cadence where the instructor made us all shout, "USA!", and then he would say, "IT'S WHAT WE BLEED FOR!"

Then we'd again shout, "USA!", and he'd scream, "IT'S WHAT WE DIE FOR!"

"USA!"

"IT'S WHAT WE'RE HERE FOR!"

"USA!"

After the run, we were marched to a hangar-type building

for graduation. The graduation was much smaller with significantly less people than the graduation for bootcamp on Parris Island. Some of the Marines' families had come to support them and I envied them since mine could not be there. Graduation happened to be on a Friday, so once we were done being introduced to our units, everyone could go off base and spend the weekend with their parents, friends, and family. Because my own parents could not be there, I knew that I was going to be left on my own all weekend as a boot in my new unit. That could spell bad news!

After the graduation ceremony, I remember feeling somewhat depressed as I saw everyone reconciling with their families. Granted, not everyone's parents came but there was still a good portion who did. I felt that ominous feeling again that I had felt the day I graduated from bootcamp, only I felt it even stronger. Whatever my gut was trying to tell me was clearly getting stronger. Whatever it was that was lurking on the horizon, whatever bad thing was going to befall me was imminent and I knew it, I just couldn't put it into words.

After the ceremony, we walked back to the barracks in small groups. By that point, the combat instructors had essentially disappeared, and I never saw any of them again as their job with us was finished. I got back to the barracks and began to gather up all of my gear to stage outside. We had to stage it all outside because Marines from my unit were coming over in just a short time to collect us and it was obligatory to be where we were supposed to be when we were supposed to be there. As I was gathering up my gear, Hyzne approached me, and we reconciled from the petty fight we had a few days prior.

"Verlice," he said with a look of sincerity on his face. "I like you. You're a good kid. It's not often that you find someone you can relate to." I smiled and responded, "I like you too."

We shook hands.

"Too bad you can't come with us", I joked, letting out a chuckle as I picked up my seabags. I'll never forget the look he gave me; he turned very serious, the smile slowly disappearing from his face and then he muttered, "Verlice, there is no force in hell or on Earth that could get me to go to [The Unit]".

I remember his reply hitting me strangely because, during the argument we had earlier, he had been going on and on about how he was jealous of me being active duty and how he would give anything to be stationed in an active-duty unit. It was weird to me that he would change his mind all of the sudden on the last day of ITB. We shook hands again after he gave me his phone number for me to contact him later and then he left out of the squad bay. I never saw or heard from Hynze again.

Shortly after, I collected my gear and took it outside to stage it in the group I was told earlier that morning to stand with. There were about 10 of us in the group and one of them was that kid, Drummer, who I had disliked so much. Essentially, they were beginning to break us off into what company within the unit we were going to be in. So, of all the rifleman in my ITB training battalion, only 10 would be in the same company along with me. Castillo was also there and a few other guys including one guy from my bootcamp platoon. My buddy, Holdin, was going to be in another company and it turned out that I never saw him again.

It's funny the way the Marine Corps works; one day you're tight, the next day they're gone. It's almost like they were always getting us ready for combat; getting us ready for the day that we may lose our buddy in battle.

By that point in the year, it was beginning to get hot outside, and I dabbed sweat off my brow as I waited for the Marine from my new unit to come get us and take us over to Camp Lejeune. After waiting for what seemed like hours, we finally saw several large 10-ton military troop trucks pulling in and parking. A few Marines got out dressed in the dress uniforms known as the "chucks". They were the khaki short sleeve dress shirt with green pants and black shoes.

All the Marines who disembarked from the back of these troop trucks appeared to be Sergeants. I'll never forget the Sergeant who came over to greet the 10 Marines, including myself, who were waiting to be collected. He looked like the real deal; he had a sleeve tattoo protruding from his short sleeve dress shirt which went all the way down to his wrist. Flames and bullets, I remember. The guy looked like a Marine; he looked like he had probably killed a few people in Iraq, for sure! I remember that being one of the most prevalent thoughts in my mind as I saw him striding over to the ten of us standing there waiting.

He introduced himself as Sergeant Howitzer and told us to, "Get the fuck in the trucks with your gear. Quickly!" He did not seem like the friendliest of guys and bore a grimacing look of irritation on his face the entire time he spoke to us. I couldn't help but notice the ribbon stack on his dress shirt. It was huge! The guy was seriously decorated. We all yelled, "Aye, Sergeant!" and then grabbed our gear, threw it in the back of the trucks, and climbed in. After a short time, he also got in and the truck began to pull away, taking us to our final destination: Camp Lejeune.

CHAPTER 6

WELCOME TO THE FLEET, BOOT!

CAMP LEJEUNE

On the drive over to my new unit, I remember all the new infantry Marines – now called boots – were sort of huddled together talking amongst each other. There was a lot of "bum scoop" (bullshit talk/ rumors) going around between us and everyone was pretty much on edge. I overheard some of the boots saying things like, "oh, I hear that they're going to do this to us…", or "I hear they are going to make us do this…", as well as other fear-laced fantasies.

One of the Sergeants who was sitting towards the front whispered something to another Sergeant and one of the boots who had overheard muttered under his breath, "Man, they just said they're going to lock us in a room and make us fight with each other." The secret banter by the Sergeants had my thoughts ablaze with all kinds of horrors about the hazing which I knew I was going to be subject to as a brand new boot fresh out of infantry school.

It seemed to take forever to get to our destination even though Camp Lejeune was only a few miles from Camp Geiger. We drove through the gates, and I remember having this feeling of...*well, here I am* as the truck pulled through the front gates.I had arrived in Hell!

Soon the truck drove us up to where the unit was located – right outside the company office. I ended up finding out on the way to the office that I was going into Alpha Company. My Unit was comprised of several companies: Alpha Company, Echo Company, Fox Company, and a headquarters company (HQ Company).

Alpha, Echo, and Fox Company were what they called the "line companies" and consisted of all infantry Marines. The HQ Company, which was attached to My Unit, consisted of non-infantry Marines such as communications Marines, logistical Marines, and any other type of Marine which supported an infantry function. There were only males my unit during the time I served as females were not allowed in the infantry.

Within each infantry line company there were four platoons – 3 rifle platoons and a weapons platoon. Alpha Company had 1st platoon, 2nd platoon, and 3rd platoon which were rifleman platoons, and then weapons platoon made up the 4th platoon. As it was in ITB, weapons platoon was made up of machine gunners, mortarman, and assualtmen. I would find out later that day that I was going to be Alpha Company 3rd platoon, Once again…"turd platoon"!

Immediately upon stepping out of the troop truck, all the boots were ushered into the company office where we met our CO (company officer) whose name I cannot remember as he did not remain my CO for long. I remember he was a Captain, a bit on the heavier side, and looked a bit disheveled overall – sort of like he had seen one too many things in "his day". He told us a few things about the unit, and I remember him saying, "you should be proud to be here. Not a lot of men can make it this far. You guys are the athletes of the Marine Corps!"

He was obviously trying to juice us up, but he was right! Besides the special forces, the USMC infantry was the hardest thing one could do in the military, and I was quite satisfied with myself in that moment to be standing at the position of attention listening to my new CO congratulate us on a job well done.

He went on to inform us that there was a lot we were going to have to take in now that we were infantry Marines in the fleet and that we should learn to listen closely to guys who had been in longer. Shortly thereafter, the ten of us were taken back outside where we were lined up right alongside the company office.

We were awaiting the arrival of a Marine who would also oversee us in Alpha Company – an enlisted Marine – my company First Sergeant, First Sergeant Kellie. First Sergeant Kellie would become one of the main aggressors later in my USMC experience and would ultimately become the bane of my existence.

I'll never forget him coming out of the company office to

greet us that morning. He was walking hurriedly while taking quick and furious puffs of a lit cigarette which dangled from his pursed lips. As we stood there, I was temporarily blinded by the sun gleaming off the black asphalt, but I could clearly see him throwing his cigarette onto the ground as he set himself up in front of us. He didn't say a single word but just looked us up and down as if we were precious cargo he was inspecting. He reminded me of a mix between Bruce Willis and Brad Pitt. He sported a small, strictly-in-regulation mustache which was almost as proper as his overly pressed and starched blouse top.

After a few short moments, he let out an overly exaggerated sigh, shook his head, and hocked up a loogie, spitting it on the ground before saying the only words he would speak that morning.

"Alright! Let's go kill some *fuckin'* ragheads!"

We all looked at each other and thought the same thing: *did he really just say that?* As an infantry Marine, that was the dream; we had trained for it, sweated for it, bled for it, and cried for it. I didn't know if any of the training was going to pay off, but one thing was for certain... I wanted to kill a "raghead"!

After the quick meeting with the First Sergeant, the famous "dog and pony show", for which the Marine Corps is notorious for setting up, began with all its miraged glory. All the families which attended the ITB graduation were there, and they sort of had the area in front of the company office set up like a mini barbecue. They were really trying to sell the "closeness" of the unit by acting like we were one big family. It was all smoke and mirrors though as it was the only time I ever saw any type of display like that during the rest of the time I was in the Marines. If there were families around, the unit would be on their best behavior - nothing but smiles, laughs, and slaps on the back.

I remember standing around listening to various speeches by the BC (battalion commander), the company Sergeant Major, and several other Marines, including the Sergeant of the guard, within Alpha Company. They were assuring all the families that their new infantry Marines would be safe, and no harm would come our way. They told them, "Your Marines are in good hands, trust us". Honor, Courage, commitment and all that bologna.

I looked around and saw other senior Marines within the company just standing around eyeballing the area. Those Marines were what they called the Senior Lance Corporals, and they were the backbone of the Infantry. Even though they were only two ranks higher than me, they were God; and they were mean! I could practically see some of them frothing at the mouth, eager to get the boots alone and start the hazing.

I'm sure that the whole company was briefed that morning before the "boot drop" and were most likely told not to touch anyone, mess with anyone, or do anything that will shed a negative light on the unit while families were present. If it was one thing the Marines hated it was negative publicity.

As I was standing there, a Lance Corporal approached me slowly, looking as though he wanted to chew my face off. I had been eyeing him from my peripherals, and he had been mean-mugging me for some time. I wasn't sure what to do or how to approach the situation, so I just stood there. As he came closer, I could see hatred in his eyes – a true boot hater and there I was… the boot! He whispered in my ear as he passed by me, "Fuckin', boot. We're gonna kill you later, Bitch!", and then disappeared into the crowd somewhere. I remember thinking about how long a day it was going to be if me just standing there warranted that type of threat.

After the speeches, the ten boots and I were rallied again, this time by a Corporal whose last name was Jacobs. Corporal Jacobs told me that he was going to be my squad leader. I was going to be in 3rd platoon, 3rd squad, 1st fire team. Each platoon was broken into squads which were then broken into three different fire teams. A fire team consisted of a fire team leader, typically a senior Lance Corporal or Corporal. Then there was the automatic rifleman who carried a SAW (squad automatic weapon) which was essentially a mini machine gun. The SAW gunner was typically the next senior guy behind the fire team leader. Then there was the assistant gunner which assisted the SAW gunner and who also carried an M16-A4 or sometimes the M4. The assistant gunner was the next senior after the SAW gunner. Lastly, there was simply the billet: rifleman. This was the least senior of the fire team and was almost always a boot. I was going to be the least senior rifleman of Alpha Company, 3rd platoon, 3rd squad, 1st fireteam. Completely at the bottom of the totem pole!

The other 10 Marines who arrived at the unit with me were broken up and introduced to their own squad leaders and fire teams. I was the only new boot in my fire team. The anticipation was really starting to ramp up. I mean, there I was, in an infantry unit being introduced to my squad leader. I had been in the Marine Corps for only 5 months, and I was already in the fleet in a deployable unit. Things moved so quickly that it felt as if my head was spinning.

Corporal Jacobs was a pretty decent guy, or at least he seemed that way to me. He was one of those guys that looked a lot younger than he really was; he didn't even have a single hair on his face to shave. He was probably in his mid-20s but looked like he was 19. He was a married man and lived off base, so he gave me his phone number in the case that anything "happened" to me during my first day in Alpha Company. Corporal Jacobs also told us that he had been deployed several times and based on what he was telling us, he was quite the decorated Marine for only being a Corporal.

After that meeting with Corporal Jacobs, a bunch of us boots were ushered here and there while several Corporals, including Corporal Jacobs, showed us around. I mean, there was not much to really see since the Marine Corps notoriously receives the least funding out of all the branches, meaning there were very few facilities around. They showed us the barracks we would be living in, the laundry room, a little "rec" room which had nothing more than a couch and TV, and the office on the bottom floor of the barracks where there was always a Marine on duty.

I remember walking around the barracks and seeing all the senior Lance Corporals on the catwalk of the barracks floors staring at us. Some taunted us, some spit down on us, and one even oinked. I remember another boot standing next to me said, "did you just hear that one oink?" I think we were all nervous by that point and it was only going to get worse from there.

After a couple hours of walking around and getting different lectures by different Sergeants and Corporals, we were eventually ushered back to the company office. Once back inside, we were assigned rooms to which we would be sleeping in. The way they typically assigned rooms to boots was simple: two boots in a room with a senior Lance Corporal - a total of three to a room. It basically guaranteed that we would all be hazed since the senior Lance Corporals of the company were itching to get us in their sights. They told me my room number (304) and also gave me my key – a small brass key with the words engraved: "Property of the US Government".

After receiving my key, I went back outside and that's when that kid, Drummer, approached me. "Looks like we're going to be rooming together!" he said, in a somewhat cheerful, yet goofy manner. "Yep, looks like it", I responded tersely. Truthfully, I was not that excited about being in the same barracks room as Drummer. I didn't particularly like Drummer at first and again just thought he was a douche bag. "We're going to be bringing so many women up to this room", he laughed, continuing his goofy persona. I just nodded and smiled. *God*, I thought to myself.

Why does it have to be Drummer?

Shortly thereafter, Corporal Jacobs approached me again and told me to follow him to where there was a huge battalion formation. The BC and Sergeant Major spoke to the entire unit, giving us our weekend brief and then finally dismissed us. After the battalion formation, Corporal Jacobs found me and then walked me to the barracks to show me where my room was. He told me once again to call his phone if I encountered any problems and also ordered me to call him the following morning whenever I woke up. I wasn't sure why he wanted me to call him first thing in the morning, but I'm assuming it was because he wanted to make sure that I was not assaulted or hazed at some point during the night. Perhaps he was ordered by someone to make sure of that, who knows? I agreed to his demand and then he walked away without saying anything else. That was it… I was on my own from there on out.

By that point in the day, all the families who had attended the ITB graduation and the company "get together" began to leave with their boot Marines. Most, and I mean almost all, of the other boots either accompanied those families and their friends for a weekend off base or took off in groups of two and three simply just to escape the hazing we all knew was coming.

As I stood outside barracks room number 304, I felt my heart begin to beat rapidly. I knew that there would be a senior Lance Corporal in the room, and I could only imagine how the first meeting was going to go. It was the first time I was left to my own devices as a boot in my new unit and I was really expecting the worst. I took a deep breath and opened the door.

THE FIRST WEEKEND

Once I entered the room, I immediately saw a senior Lance Corporal who appeared to be packing up his things in a very furious manner. He was a somewhat stout guy with a low-fade hair cut – the hallmark of a senior Lance Corporal – and was clad only in his green skivvy shirt and Cammie trousers. He appeared to be maybe 21 or so and I guessed he was of some Hispanic ethnicity given his darker skin complexion. He seemed to be packing all his belongings and appeared to be very angry, huffing and puffing the whole time as he haphazardly threw objects into his bag. Leaving for the weekend, perhaps? – I wondered. He immediately saw me standing in the doorway with all my seabags and body bag which held all my uniforms and darted a glance of disgust at me. Obviously, he could tell I was a boot not only because of my youthful appearance but also because of my shaved head.

He instantly stopped what he was doing and approached me and barked, "What the fuck are you doing in here, boot?" I began to stammer, "Um... I was assigned to this room." He appeared to become even angrier with my response. "Uh...I know we say FUCKIN' AYE LANCE CORPORAL!", he shouted loudly.

It was a bit strange to me that he wanted me to address him by his rank as it was told to us over and over in ITB that we did not need to address Lance Corporals by their rank since they were non-commissioned E-3s. I realized at that point that it was obviously a lie told by the combat instructors to mess with us, and I found out right in that moment that all Lance Corporals were to be addressed by their rank – no different than a Drill Instructor.

"Aye, Lance Corporal!", I responded sharply. "Yea, no shit, bitch!", he screamed back as he eyeballed me with pure distain. I was so uncomfortable, and I had no idea what I was supposed to do, so I just continued to stand there like a deer in headlights. He started grilling me and quizzing me on a bunch of stuff, like Marine Corps knowledge, and was almost trying to act like a drill instructor, which I found incredibly annoying. I was thinking to myself, *Bitch...I'm not even trying to deal with this or be treated this way by someone who is only a couple years older than me.*

Age, however, did not matter in that situation; it was all about experience and because all the senior Lance Corporals in Alpha company had been to Afghanistan once, they were basically God to a boot like me. We could not touch them, we could not fight them – lest we get ganged up on, and we could not argue with them. To a boot like me, they basically held the same power as a Drill Instructor.

He continued, for about 30 minutes or so, grilling me on various Marine Corps history and other rifleman knowledge that I "*should* know" from bootcamp and infantry school. "You were just there, boot. You should fuckin' know this shit!", he would say if I didn't know the answer to one of his questions.

I distinctly remember him asking me to call out a 9-line. A 9-line was something that would be called out over the radio in the event of someone being shot or wounded in combat. It was essentially how one would call a medivac to a specific location for evacuation. The 9 line was something that I learned in ITB but for the life of me could not get right that day as he was grilling me.

Eventually he stopped asking me questions and continued with his packing, albeit even angrier than he was before. I awkwardly set my things down on one of the bunks and slowly took a seat, feeling my heart pound in my chest as I tried to relax. I observed the room and saw that it had three beds: a bunk bed – top and bottom bunk – and one single bed. The single bed was reserved for the most senior guy in the room leaving the bunk beds for the two boots.

As the senior Lance Corporal continued to pack, I could

hear him every so often whisper under his breath, "fucking bitch" or "boot mother fucker!". It was like, damn, dude! I literally did nothing more than just walk into the room which I had been assigned, and the guy was acting like I took a huge shit all over his bed. It really didn't matter; the fact that I was a boot, I guess, was crime enough.

The room, I remember, was God-awful with an almost eerily uncanny resemblance to a prison cell. The place literally looked like something which should have been condemned and looking back, I'm sure it probably was. It was dark, drab, cold, with cinderblock walls and weird hard epoxy-type flooring. In addition to the three beds, there were three large wall lockers to keep our gear and clothes in. They were the same exact wall lockers that they had at Camp Geiger and I'm guessing every base in the Marine Corps had the same ones. The sheets to the bed were also the exact same as well.

Can we have a bit of variety for Christ Sake?

I remember everything in the room being broken or dented, beaten or bashed, and basically all of it wasn't suitable for even the worst convict - in my opinion. The sink was situated outside the bathroom door the way they are in really shitty motels, and the bathroom itself was so small that one could barely fit in comfortably. The whole room was just a bit bigger than a prison cell and I concluded, again, that it was hardly livable. A common theme for Marine Corps ammenities.

I sat on the bottom bunk as Lance Corporal Yimudio, as he would state his name was, continued to throw things haphazardly into his duffle bag. As I sat there, uncomfortably, another senior Lance Corporal walked into the room and introduced himself as Lance Corporal Dudley. Lance Corporal Dudley was one of those guys that I could tell was not the sharpest knife in the drawer. He seemed to be from Arkansas or Alabama somewhere and I could almost picture him with a huge buck wheat weed dangling from his mouth as he spit chewing tobacco on the ground. He seemed like a nice enough guy, though, and reminded me of Forrest Gump.

I remember I was looking at my phone at a picture of my girlfriend, Gabby, and Lance Corporal Dudley saw it and smiled, saying, "is that your girl?" I looked back, smirked and said, "yea!"

Lance Corporal Yimudio heard that I didn't respond with the proper "rank" and screamed right in my face, "BITCH, I KNOW WE SAY AYE, LANCE CORPORAL! HE IS A FUCKING LANCE CORPORAL!" I agreed and corrected myself by responding with the proper rank, "Aye, Lance Corporal!" to Lance Corporal Dudley who just rolled his eyes. It was a power trip to the max! Lance Corporal Dudley whispered in my ear, "don't worry, he's a dick!" and then proceeded to stick his middle finger in my face as he walked out of the room giggling.

JUMPED!

Eventually Lance Corporal Yimudio left the room, and I was left alone, which was a welcome change for me. The only thing I wanted to do then was get off the base and as far away from those senior Lance Corporals as I could because I knew each one of them would be out looking for blood that night. I quickly changed into my civies (civilian clothes) and exited the room, checking over my shoulder every few moments to make sure I wasn't followed.

I made my way down a street and started looking for a taxi that could take me off the base and into town. While I was waiting, I ended up running into a friend of mine from bootcamp who had also been assigned to the unit. We both agreed that we wanted off the base and found a taxi to take us.

The taxis in Jacksonville, North Carolina preyed on the Marines stationed at Camp Lejeune. As a matter of fact, most people out in town would prey on the Marines in any way they could for whatever reason – usually in a scam artist-type way. It was easy to spot us, obviously, and I swear they would swarm us like vultures. They would try and sell us bullshit we didn't need or scam us in more than a few outlandish ways. Over the next year, I became extremely sharp at spotting a potential scam or predator looking to do me harm.

My buddy and I had the taxi drop us off near the mall where we ended up getting something to eat at one of the restaurants. Afterwards, he decided that he wanted to watch a movie at the movie theater which was connected to the mall. For whatever reason, I decided I didn't want to go in and watch it and told him I would wait outside for him and reconnect whenever it was over.

I waited several hours and after the movie was over I looked around for but never did find my buddy. I'm not sure where he went or what happened, but I'm assuming he just figured I went back to base and decided to deuce out. So, there I was, alone out in Jacksonville, North Carolina – a boot all by myself. No part of me wanted to go back to the base mainly because I was afraid of being hazed or messed with by Lance Corporal Yimudio whom I assumed was back in the room by that point.

By that time, it was probably 9pm or so and I figured the bookstore at the mall might still be open. I was planning to just spend as much time there as I could until exhaustion forced me back to the base and then, because I would be so tired, perhaps I wouldn't care about being beaten up by Lance Corporal Yimudio. I began walking away from the movie theater and towards the other end of the mall where the bookstore was but in order to get there, I had to traverse a rather dark and abandoned section of the parking lot.

As I walked through the empty parking lot, I noticed off to my left a large group of about 7-10 guys. They appeared to be civilian guys – black guys – and appeared as though there was possible gang activity going on. I tried to hurry along but they quickly spotted me before I could clear their line of sight. I wasn't paying too much attention like I should have been, but one guy approached me head on and began walking backwards, matching my pace, and asked me for a cigarette. I told him that I didn't have one and he asked me several more times. What I didn't see was the other group of guys swarming behind me getting ready to attack.

Suddenly, I was punched from behind by one of the other guys who had flanked me. All hell broke loose as I was essentially punched everywhere on my body: ribs, back, head, face – everywhere. There was no way to defend myself or even focus on one because every time I tried, another punch or kick came from another angle. Eventually, I did the only thing I could and that was get into a crouch position with my hands covering my head for protection. They continued to punch and kick at me for several more moments until they finally all peeled off and disappeared as quickly as they came.

I was in sort of a state of shock and my adrenaline was pumping so hard that I didn't even feel the pain at first. I quickly shot to my feet and looked all around me to see that they were gone. A lady in her car drove up and asked me if I needed an ambulance or the police to which I declined. I did NOT want my command to find out about it because somehow I would end up in trouble for being alone out in town. It was my first day in my unit and already I had been jumped and that was not something I wanted other Marines in my new unit to be aware of.

I told her I was okay, and she eventually drove away. I walked, casually, to the bookstore where I had intended to go all along – just as if nothing had happened. Once in the store, I walked into the bathroom, mainly to check my face and make sure that I wasn't bleeding and it was then that the gravity of the situation gripped me.

As I looked at my face, I saw no blood, but I did see bruises all over. I had a black eye, my forehead was bruised, and my cheeks were blue. My ribs and back were sore as well. *Jesus, how would I explain this?* I thought to myself. The adrenaline began to wear off and I started to shake, almost going into a state of delayed panic. I was lucky that they had not stabbed or shot me but honestly, I was more worried about explaining the bruises to my command if someone was to start asking.

Eventually, I ended up getting a cab and going back to Camp

Lejeune. I was too tired to keep walking around aimlessly and honestly, the incident I had just endured had me a bit shaken up to the point where I didn't care what Lance Corporal Yimudio might do to me back in the room. Once I got back to base and to my barracks, the phony tough guy bravado wore off and I became nervous to go back into the room again. It sucked being a boot and was almost as bad as being a recruit, and if I thought getting jumped by a bunch of hoodlums out in town was bad, it would be nothing compared to what a few senior Lance Corporals would do if they got their dick skinners on me.

I walked around the catwalks of the third deck, circling it over and over again aimlessly. Every time I passed my room, I would listen closely to try and hear some noise or any signs of movement. Last thing I wanted to do was burst into the room while Lance Corporal Yimudio was sleeping or worse…in there with a couple of his buddies. That would be a nightmare for me, and God only knows what would ensue! I continued to walk around some more and eventually went downstairs to smoke a cigarette at one of the smoke pits.

As I was standing there smoking, I saw a couple of Marines walking by. I could tell just by their haircuts that they were senior Lance Corporals. Of course, they were able to tell that I was a boot just because I looked young, and my head was shaved. They both approached me, and I could tell that they were both pretty drunk. One of them, a smaller blonde guy, approached me and got into my face. He started to threaten me, grab me, and mess with me saying that he was going to kill me. For what? Who knows, I mean I was a boot, so I guess it didn't matter. One didn't need to do anything as a boot for a senior Lance Corporal to want to kill you. Just being a boot was enough. I remember just saying "Aye, Lance Corporal" over and over until they eventually went away.

I went back upstairs and decided to hide in the laundry room of the third-floor barracks. It was pathetic! There I was, supposed to be a tough Marine, and I was hiding in a laundry room like a scared little kid. I remember hearing the wind whip through the open hallway, creating a ghostly whistle which sent a cold shiver down my spine. It was then that I finally decided to grab my nuts and be a man and head back to the room to face whatever was waiting for me.

I walked back to the room and just opened the door slowly, hoping not to wake anyone who might have been in there and, to my surprise, it was empty. I had no idea if Lance Corporal Yimudio was returning, but by that point I didn't care. I was too tired, and too much had happened already. I just lay on my bunk, fully clothed, and slowly fell into an unsatisfying sleep.

The next morning, I awoke and was surprised to see that nobody had entered the room which I had brazenly left unlocked. I had half expected Lance Corporal Yimudio to bolt in, wake me up, and start hazing me but luckily, he, nor anyone else, did. I remember checking myself in the mirror and seeing how swollen and bruised everything was because I was feeling the pain in both my face and body by that point.

I was worried about someone asking me what happened to my face but decided I would cross that bridge when I came to it. I still didn't want to be hanging around the barracks since I was likely one of the few boots around, so I just left the room and headed away from the barracks and towards where I knew the shops were located.

I remember walking over to the large MCX (Marine Corps Exchange) store and there I just sat in a Gazebo near the loading dock of the complex. I essentially spent most of the weekend in that area just smoking and keeping low. I remember feeling completely out of place, lonely, depressed, and home sick. I did end up calling Corporal Jacobs, as instructed, but did not tell him anything about the incident out in town the night before. I figured if someone asked me, I would just make up a story. Nobody really asked me about it, and I eventually healed up.

MY UNIT

So, there were some very interesting things about my unit which I believe ultimately led to the chain of events which sparked the macabre assaults, sexual harassment, and mental abuse which ultimately befell me in the Marine Corps. I truly believe that if I had not been put into that corrupt unit, I would be telling a very different story right now. My Unit was corrupt and that is all I will say about it for anonymity and other personal reasons. It may sound dramatic, but it was objectively true given the activities the unit had been involved in "historically".

The USMC was in the process of gutting the unit's chain of command when I arrived to "promote" a more hospitable environment for the Marines attached to it. I had, by pure bad luck, walked into a brush fire from which there was no escape. It was something no Marine wanted – to be stuck in a bad unit – and

163

was something that no Marine had any control over. The Marine Corps had a strict obligation to the mission and we, as Marines, went where the Corps needed us and that was that.

If the Corps put a Marine into a horrible and corrupt unit, then that Marine was going to stay in that unit come hell or high water no matter what. It didn't mater how much bitching, whining, crying, begging, or reasoning one was capable of. The Marine Corps gave zero fucks and if, and only if, a Marine made it through the first four years then just *maybe* they would be able to move to a different unit. That was my experience anyway.

SETTLING IN

It turned out that Lance Corporal Yimudio never returned to the room and in fact had moved into another room with one of his senior Lance Corporal buddies. I guess he would've rather stay with one of his peers than with a couple of new boots. By some fluke, my room ended up being only assigned to boots. I guess it had been some type of overlook by command and the mistake was never caught.

Since we had the only room without a senior Lance Corporal, the place became a safe haven for the other new boots, and they would regularly knock on our door to hide and avoid hazing by the seniors which they were roomed with. Drummer and I became very close friends during the time we were in the barracks together and would often go out into town, eat at the chow hall together, and just goof off whenever we had down time.

Still, Drummer was always a bit of a douche bag, and he would never shy away from an opportunity to make me feel like shit about myself, poke jokes at me, or be overall crude around me. He would walk around the barracks room naked, making jokes about his dick, flapping it around and stretching the skin of his ball sack out so it resembled something he dubbed "the bat wing".

Sometimes he would yell for me to come and look at the nasty shits he would leave in the toilet or show me the random loads he would shoot in the bowl after masturbating. Honestly, it was pretty common behavior among the guys in the infantry and it became clear early on that nothing was sacred; we got to know our buddies quite well! Looking back on it, it grosses me out, but at the time it was funny as hell. We would laugh for hours at night talking about this and that, prank calling people, and just being goofy 18-year-old idiots. It was a little bit like what I imagined summer camp would have been like.

My other barracks mate was a more "senior" boot Marine, who had been in the unit about two weeks longer than Drummer and I, and his name was Mitzner. Mitzner was a bit of a wimp, and we all thought it. He was very timid, awkward, and a bit of a loner. He would not participate in the fuckery Drummer and I would partake in and usually made himself scarce on the weekends. He did have one friend that I remember whose last name was Carrol. Carrol was even wimpier than Mitzner and he was also the biggest mama's boy I had ever met. Both were a bit odd and neither Drummer nor myself liked either one of them. Both Mitzner and Carrol ended up deserting later and were both ultimately discharged on unfavorable grounds as a result. Birds of a feather flock together, I guess.

That first week in the fleet was where I was introduced to everyone who would be in my fire team. My fire team leader was a Corporal whose last name was Richards. Corporal Richards had been in the Marines for roughly 6 years or so and before being infantry he had driven trucks with Motor T (Motor Transportation). He had made a lateral move to 0311 after his first tour of duty. Typically, a Corporal who was not infantry by trade, like Corporal Richards, was not as respected by the other 0311's, especially the senior Lance Corporals, and even though he outranked them, it was unspoken Marine Corps law that he was not as high on the totem pole as they were since they had done a rotation in Afghanistan as infantry Marines already.

Regardless, Corporal Richards had been to Afghanistan and was not as "boot" as some of the other Corporals who came from Security Forces. The Security Force Marines were the worst! They signed a five-year contract where their first three years consisted of doing security details at Guantanamo Bay and other embassies. After their first three years they made Corporal and then rotated to the fleet to finish their last two years in an infantry battalion, usually being given a fireteam or squad leader billet based on their rank.

The security force guys were basically glorified prison guards! They had no combat experience and were all around hated by the senior Lance Corporals due to their lack of experience and the ease of which they picked up the Corporal rank. Even though the security force guys had no combat experience, that did not stop them from being the biggest cock suckers to ever plague the Corps. They almost seem to try as hard as they could to be tough and make the boot's lives as miserable as possible. Even those jerks, however, were not respected by the boots the same way as the senior Lance Corporals who had combat experience were, even though they technically outranked them. Every Marine in the infantry referred to the security forces as "Se-Queerity Forces".

Corporal Richards was basically a dick head; he would take every opportunity to fuck with me, make me feel like shit, haze me, or just taunt me. I remember the first time I met him that week, he screamed at me for not addressing him as a Corporal and almost threw me through a wall. Many bad things ended up happening to Corporal Richards and he did not end up being my fire team leader for very long as a result. The other two people in my fireteam were senior Lance Corporals: SAW gunner – Lance Corporal Chance and A Gunner – Lance Corporal Ceitur. Both were pretty decent guys and didn't mess with me too much. I, of course, was the boot rifleman.

The first two days I was in the fleet were pretty busy and consisted of a lot of shots and medical stuff similar to how it was during the first couple days of bootcamp. They pumped all the boots full of random shots which I am still unsure of. Who knows what they were giving us? Flu shots, typhoid shots, tetanus, malaria? I mean, who the hell knows – could have been salt peter as far as I was concerned. We also went to CIF which was a place where they issue all the combat gear such as our packs, helmets, flaks, grenade pouches, bags, and everything in between. We were also issued our rifles from the armory – of course for me it was the M16 A4 Service Rifle…my old friend.

It was also during that week that I met my Platoon Sergeant, Sergeant Walker. Sergeant Walker was not only one of the meanest Marines I ever met during my tour, but one of the meanest people I ever met in my life. The guy was downright evil, but most importantly he was experienced and that was really all that mattered. He had been to Iraq and Afghanistan multiple times and had even won the Silver Star medal for gallantry in action. He was so mean that even other Sergeants didn't mess with him and the command gave him a *very* long leash to play on. He only hung around a select few guys and never comingled with any of the boots or even most of the senior Lance Corporals.

The first time I met him, he barged into the room and screamed, "I'm Sergeant fucking Walker, do you understand!?"

"AYE, SERGEANT", we all screamed back. He made us all stand at the position of attention almost like we were back in bootcamp which told me a lot about what type of leader he was going to be. He was honestly a scary guy and one of the few guys who truly put fear into my body. If I saw Sergeant Walker on the streets today, I would spin on my heels 180 degrees and run in the opposite direction. No joke!

With the first couple days out of the way as a new infantry fleet Marine in a deployable unit, the only thing left was to train for deployment. And train we did! We trained very hard as a matter of fact. The training we did was harder than bootcamp and infantry school put together and made both look like a cakewalk!

Every unit is different but because my unit was corrupt and constantly getting a new CO, we were consistently changing our pace of training from hard and harder to fucking impossible, and even though we weren't considered special forces, we trained like it. It was very common for us to partake in 12-mile pack runs,

6-mile Indian Sprints, 8-mile log runs, and every other type of workout imaginable - everyday.

TRAIN HARD, PLAY HARD

So, we began to train every single day while in garrison (home base), and they did not take it easy on any of the boots. The best way I can describe the level of physical training and mental exhaustion is to compare it to the crucible. Life in the unit was almost like the crucible every day. We were constantly under mental and physical stress to the point where some people would start to go mad and have little breakdowns. During the first couple weeks in the fleet, many of the boots, including myself, were wanting to run away and go home. Between the constant hazing every afternoon after "work" and the incessant physical training, it was almost too much for a person to handle.

The camaraderie that the Marine Corps boasts itself on fostering appeared to be nothing more than smoke and mirrors to me – a marketing gimmick to get young, naïve kids to join their "club" for a phony cause. The senior Lance Corporals wanted nothing more than to kill one of the boots, and I mean literally! Some of them, if given the chance, would have taken and eviscerated us with a smile on their face. Granted, there was some "camaraderie", but it only existed between the boots. We were all going through a "shared hell" together and it did cause some of us to grow a bond together, stick close to each other, and watch each other's back, but still; it was not what I had expected. I had expected to be treated like a "new Marine" but also like a member of the club. It was not like that at all!

A typical morning in the fleet began at 6 am, or 0600 in military time. We would wake up, get in our PT gear, which was just the green skivvy shirt, green skivvy shorts, and go-fasters, and then go outside right in front of the barracks to get into formation for the morning counts. After our platoon Sergeant or commander made sure we were all there, they would break us for morning PT. The morning PT would be done in various ways depending on the plans of the platoon Sergeant or the Lieutenant. Sometimes it was just PT in our fire teams; sometimes it was squad PT; and sometimes it was platoon PT. Occasionally, we did battalion PT with all the other companies, but that was a rare occasion.

For me personally, fire team PT and squad PT were the most physically demanding because they were in smaller groups. It would be up to the fire team leader or squad leader what we were going to do, and it was usually something meant to completely break us off. The PT would usually last two or three hours and would almost always include a run of some kind. One thing I noticed in the fleet was that they didn't give two flying fucks how fast anyone's run time was. It became apparent to me that I was to keep up with the fire team or squad even if I was incapable of doing so.

To make matters worse, the fire team leaders and squad leaders were all "PT gods" and could easily run a 16-minute three mile without batting an eye. I was just not capable of running that fast; and it wasn't just me. Many other boots were not able to keep up with them either and almost everyone would be broken off on a daily basis as a result of the morning PT sessions. My 3-mile run time was about 22 minutes, which was still not slow by any means as they gave you 28 to do it in, but that didn't matter. To them I was the slowest mother fucker who ever served in the Corps because I couldn't keep up with their pace even though I was well within the PT run time limits.

At least in bootcamp they tried to put everyone in groups of others who had similar run times and therefore had a chance of keeping up with each other. In the fleet – in an infantry battalion – they did not give a fuck! It was not a matter of sink or swim; it was swim or swim, fuck if you know how to or not! They would run me to the point of near death with 12-mile "gate" runs, Indian sprints, log runs, pack runs, 8-mile runs in a gas mask, you name it. Every day was something different and every day I didn't know if I would be able to go one more.

Personally, I did not see the value of being run to the point of being broken or half dead every single day. There was no training value in to be gained from it because it leaves the body open to injury which directly contradicts the Marine Corp's purpose. Why run someone to the point of injury? It takes an able-bodied Marine out of the fight! I saw it happen with many boots: leg injuries, shoulder injuries, foot injuries, as well as countless mental breakdowns. They would then be put on light duty and more often than not discharged medically as a result. It was stupid to me and made no sense.

To be fair, training was not really the point of the PT sessions. To the fire team leaders and squad leaders, PT was simply meant to destroy the boots, both physically and mentally. They knew damn well that I could not keep up with their stride, but they did not care. It was done for no other purpose than to just make me feel inadequate and "less than" a Marine. They would always use the opportunity to call me a "faggot", piece of shit, traitor, and everything in between! All because I couldn't run at a dead sprint for 12 miles straight – I mean who could? The only thing that I could do was try my best – to which I did – and keep going another day.

We would also train in various strengthening activities as well such as performing conditioning humps which were significantly longer than any hump I ever did in bootcamp or infantry school. We also practiced various war fighting techniques including shooting the weapons, dry firing the weapons, speed reloading, and of course cleaning the rifles for hours on end. We also learned various infantry tactics which were not taught in the infantry school such as finding booby traps, IEDs, taking prisoners, and all that fun stuff.

Every day at about 4 or 5 pm, we would form up outside of the company office where we would be given a brief for the evening on what to expect the next morning (i.e. what time to come down to formation, PT gear, plan of the day etc.). After the brief, they would break us for the day, and we could do whatever we wanted...sort of.

Unfortunately, as a boot, the day usually would not end with evening formation because we would often be caught by a senior Lance Corporal and ordered to do some ridiculous task or duty. Sometimes they would just be looking for a boot to haze and we learned very quickly that as soon as evening formation was over, running to the barracks, throwing off our uniforms and putting on our "civvies", and proceeding to run away from the barracks was the best way to avoid being caught by one of the senior Lance

Corporals.

Drummer and I got really good at avoiding their detection and would sprint to the barracks and lock the door once inside. I can remember many evenings literally hiding in my wall locker to avoid being grabbed by a senior Lance Corporal. Sometimes they'd even break into the barracks room and get us. Sometimes we would get lucky, and they would go away to find someone else. It was not the best of times for me and often I found myself at odds with my decision to join the infantry.

I know it may sound like I am bitching or complaining about all this, and I guess to some extent I sort of am; however, at the time, I never did. I never questioned any of their orders no matter how bad or stupid they were. Whatever they wanted me to do, I did and as a result, I was hazed quite a bit during those first few weeks as a boot in my unit. Sometimes it was physical and other times it was mental – something to embarrass or make me feel bad about myself. Sometimes they would come into the room at 3 am and order me to put on my uniform and go outside. For what? Whatever they wanted! Other times, they would come into the room and make me do incredibly humiliating and embarrassing things.

I remember one time a corporal and several other senior Lance Corporals coming into my room extremely drunk. It was always the worst when alcohol was involved! I remember they made me get on the floor in a pushup position while they spit on me and threw beer bottles at me. They called me all sorts of degrading names and then made me hump the floor while they taunted me saying things like, "yea, you like that don't you…feels good right, boot?"

The only thing I could do was yell, "AYE, Corporal!", and keep doing whatever it was they told me to do as it was the quickest way to make it stop. Any kind of rebuttal or defiance would just make whatever they were going to do even worse, so really it was pointless to fight it. I remember after that incident, the senior Lance Corporals sort of began to leave me alone, and outside of a few events here and there, they didn't haze me much after that.

Some of the senior Lance Corporals, the more mellow ones, began to warm up to a few of the boots after a while and even though they would still take every opportunity to make fun of me (my voice, my face, my eyes), they would sort of include several of the other boots, including myself, in certain activities.

For example, Lance Corporal Yimudio and another Lance Corporal began to ask Drummer and I to play basketball on certain evenings to which we did. There would always be a lot of "physicality" involved in the games often times with no rules – or more like prison rules - and they would always be taunting us and berating us the whole time, but it was still nice to feel somewhat included.

Not long after arriving to the fleet, I picked up rank and was then a Private First Class (PFC). I'll never forget that day because it was really the first day that I felt inadequate as a Marine. I remember during one evening formation my platoon commander, SSGT Nimbs, announced that some of the boots were going to be getting rank. SSGT Nimbs was a good guy – very easy going with an "awe shucks" kind of personality. He was from Alabama and reminded me of one of my neighbors who had raced cars on the weekends. The guy was pretty fair for a platoon Sergeant and just an overall nice guy. That evening, I remember him lining up several boots, including myself, in a formation so that he could put our new rank on our collars. I felt so good about myself, like I was advancing and getting closer to being the Marine I wanted to be.

After he pinned the rank on all the boots, we were then disbursed back to where we had been in formation originally. My fire team leader, Corporal Richards, came over to me with several other senior Lance Corporals and began to lay into me as hard as they could. I had not done anything to instigate him and for whatever reason I had been the only boot singled out when all I had done was get my rank and stand back in formation. I remember he came up to me and began to talk about how I was the biggest piece of shit the Marine Corps had ever seen and how I didn't deserve the rank I had just been given.

I was then circled by three other senior Lance Corporals who proceeded to pull me one way and then pull me another way, the same way a high school bully might do to some wimpy Freshman. Corporal Richards continued by saying that I was not worthy of being a Marine and that I was a disgrace to the uniform. He then pulled off my rank, threw it in the dirt, and stomped on it. It may not sound that bad, but considering I was the only one this happened to, it was, and it was rather ignominious. It also hit hard on another level too because I was a Marine and being called less than was one of the worst insults one could get.

what the hell did I spend the last six months of hell for if I am not a Marine in your eyes?

175

It was a crushing blow to my ego and my motivation, and the fact that I was the only one it had happened to made it that much worse. It made me wonder why I was always being singled out for no good reason. I thought back to the first night on Parris Island when that guard singled me out for my long hair. It made me feel like I was an outsider and that was not the way anyone wanted to feel in an environment like the Marine Corps.

After he stomped my rank into the dirt, SSGT Nimbs, who had seen the event transpire and obviously allowed it to happen, proceeded to call Corporal Richards up to the front where he scolded him for what he had done. After Corporal Richards returned back to the formation, he whispered in my ear, "I'm gonna kill you for that, bitch!"

I was thinking to myself, *Dude, what the fuck did I even do?* I began to hate Corporal Richards after that day. What made it all worse was that I could not touch him! I couldn't defend myself or say anything back to him. All I wanted to do was to scream in his face, "Fuck you, bitch … you ugly, bald headed fucking ass hole! I hope you step on a fucking IED and have your dick blown off!"

But you see, that was not possible. If I had done that, the punishment would have been worse than any hazing I could imagine. They would have taken that rank away from me as quickly as they had given it. All I could do was deal with it and keep my mouth shut.

The SSGTs, First Sergeants, and Officers all knew about the hazing that went on but didn't care and while they didn't "condone" it, they didn't stop it either. They all believed that hazing was an essential part of Marine Corps "training", even though the Marine Corps had a zero-tolerance policy on hazing. The much higher ups, like Colonels and Generals, all held firm that no hazing was to occur, and that was trickled down through the ranks and intended to be enforced by the First Sergeants and Captains, but unfortunately The Marine Corps, I would find out very soon, was nothing more than a giant contradiction.

After a while, I began to get a little more ballsy with some of the senior Lance Corporals and Corporals and realized that there was some pushback allowed. One night, Drummer and I went out into town and got tattoos – something every boot did at one point or another. I remember getting a really shitty, poorly done and overcharged tattoo of a half-naked girl on my upper left arm. I'm not sure why I decided to get that tattoo because I would never have gotten it had I not been in the Marines. For some reason, I thought it was so cool… like *yea, I'll be the only guy here with this type of tattoo*! It was like I was making a statement since it was a tattoo that the Marines would frown upon given its potential to be "offensive". I guess I had a bit of a belligerent streak in me by that point.

A few days after getting it, reality sank in, and I just wanted to keep it hidden and never have it seen by anyone, especially any of the senior Lance Corporals. Somehow though, a few of them ended up finding out that I had it and wanted to force me to show it to them. Basically, they just wanted to make fun of me and fuck with me over it, to which I was quickly getting sick and tired of. When they would demand to see it, I refused and most of them just called me a "bitch" and moved on. However, one day during a training exercise, a Corporal by the name of Fount ordered me to take off my shirt to show him.

Corporal Fount was one of those "Se-Queerity Force" assholes, and I really hated him particularly because of that and because of the way that he looked. He was sort of stout and nerdy looking with a little boy-type crew cut. He kind of reminded me of a fatter version of that kid Gordy from *Stand By Me*. He looked like the type of guy who probably got picked on for being a nerd in high school but could finally take out some aggression on someone lower ranking than himself.

That day was the day I put my foot down and decided I would take no more shit from those Corporals. The guy tried every which way to get me to comply, and I denied each time by telling him, "No." He began throwing stuff at me and I was getting increasingly more pissed off by the moment, but still I tried to keep my composure. I knew that a fight was most likely going to ensue and since he outweighed me by 50lbs, not including all of the combat gear he had on, I knew I would probably get my ass kicked. I didn't give a shit, though, and I made my mind up that I was going to go down fighting.

After several more taunts, he grabbed me off the ground where I was sitting and attempted to throw me like a ragdoll. That was where I drew the line! I grabbed him back as he lifted me off the ground and when he went to throw me, we both went tumbling. We fought each other for several moments and I'm proud to say that I held my own!

The fight was broken up quick and later that evening he gave me a beer and said I was "cool". Water under the bridge, I guess. Nobody ever tried to see my tattoo again after that incident and I remember feeling empowered knowing I wasn't going to be putting up with stupid shit like that moving forward.

Shortly around that same time, Drummer and I began to stop hanging out as much mainly because I was put into a different barracks room. I ended up being assigned to a room with one of the "Se-Queerity Force" guys – a Corporal. The guy was not too bad, but he snored like nobody's business and half the nights I was kept up until 2 am because of it.

I eventually linked back up with my buddy from Infantry School, Castillo, who was in second platoon, and we began to get really tight. Because he had a car, we spent a lot of our weekends going on random trips and it was so good to get off the base and away from the Marine Corps life.

Castillio was from New Jersey and one Friday asked me if I wanted to go home with him because he needed to get his license renewed and didn't want to go alone. We ended up leaving and driving all the way to New Jersey from North Carolina, getting there in the middle of the night. I remember instead of going straight to his house, he instead pulled into the parking lot of a very seedy-looking place in a very seedy part of town. "Where are we?", I asked him. "Just come on, you'll see." The name of the place, I'll never forget, was the Go-Go Rama.

Once we entered through the doors, it was clear to me what it was: it was a strip club. It was one of those clubs where the women were not of the highest quality and the characters hanging around the poles and catwalk were less than reputable. Regardless, I was excited to look around since it was the first strip club I had been to and was interested in seeing where the night would end up taking both of us.

I remember pulling out a bunch of one-dollar bills from the ATM and taking a seat at one of the tables near the front bar. I thought I was so cool with my dog tags hanging out of my half buttoned up shirt, even though I looked like I was 15 years old. I still can't believe nobody told me to get my ass out for being "underage", even though I obviously wasn't and could produce an ID if prompted. I did, however, feel a little awkward and out of place... like I didn't belong.

I remember there was a greasy-looking dude standing a few chairs away from me eyeballing the area. His hair was spiked up and he sort of had that "Jersey Shore" look going on. He was a "Guido" for sure!I kept looking at him out of the corner of my eye just because for some reason I was getting a bad gut feeling about him, like he was going to sneak up behind me and fuck me in the ass or something.

After several moments, I felt someone touch my upper thigh, dangerously close to my dick, and thought immediately that it was that Guido a few seats down. I yelled, "EYYYYY!", very loudly, only to realize it was a pretty blonde girl not much older than myself.

She looked at me up and down as if I was the sexiest thing she had ever laid eyes on – a common ploy for a stripper no doubt - and whispered in my ear, practically kissing it,

"You're so sexy!"

At that point, I was feeling confident!

"Can I show you something?", she whispered, touching my crotch. "Sure!", I sniffed, trying my best to appear confident even though I had no idea what I was doing. She spotted my dog tags hanging out of my shirt and almost seemed to be drooling.

"Oh, you're in the Army?"

The ARMY! PFFT. No, honey, I am not one of those nasty, undisciplined soldiers... "I'm a Marine!", I proclaimed with the most confidence I had ever been able to produce. She looked me up and down, "Mmm, that's hot!" She almost seemed as though she couldn't resit me or my USMC dogtags any longer and was just begging me to take her in the back and show her how the Marine Corps was different than the Army.

I wasn't stupid though and knew that she didn't really like me, not like that anyway. She simply wanted me to give her money, and yea sure, I was in a strip club alright, but ol' Chucky was cheap! I wanted to save that massive stack of ones for someone else!

"Can I give you a dance?", she whispered to me sensually.

"No, that's okay, thanks anyway!" I guess you can call me modest but the idea of some random girl giving me a dance when I looked like I was 15 years old did not appeal to me for some odd reason. In my mind, I wanted her to be my girlfriend; I wanted to talk to her and figure out her life story. I wanted to save her from the life she was leading and give her something more. It didn't take me long to realize what a dumb thought that was and I instead turned to her and questioned, "How about my friend here?" I was pointing to Castillio who was seated next to me not even paying attention. She looked at him and pouted, "but I want you!" I again declined and she eventually went away to find some other sap who would end up spending his hard-earned cash on her.

I remember spending 10 bucks on a Red Bull and then following Castillio to the front where the poles and catwalk was. They had just made an announcement that girls were going to be coming out and doing dances and we wanted to get a closer look. I'll admit that most of the women that came out on that catwalk were not what I would call "my type".

They were large and some of them were big boned or out of shape. One black woman with the biggest tits I had ever seen came out and as she was strutting around in her seven-inch heels, she spotted Castillio and I sitting there, and decided to make a b-line towards us. Our jaws were on the floor, not because she was hot but because…"God Dayummm, those tits are *huge*!", Castillio yelled, garnering several people's attention.

The giant lady approached him with about as much sass as she could have possibly exuberated and knelt right in front of him. "Oh, so you like big boobs, baby?", she cooed seductively. She then grabbed him and shoved his face into her busty bosom as he sort of flapped and floundered like a fish out of water. The look of his face disappearing into a sea of black tits was the funniest thing I had ever seen, and I began to laugh uncontrollably.

When she noticed me laughing, she then turned her attention to me. "Don't worry, Sugar. You're next!" I braced myself as she proceeded to grab my head and bury me in her chest. Gents, she was rough too! Eventually both Castillio and I recovered from our "bosom dive" and watched the rest of the women come out to do their little struts on the catwalk.

After I made it rain with my one-dollar bills all over the catwalk of the Go-Go Rama, we decided to call it a night and ended up driving to his house so we could both sleep. Castillio clearly came from a very poor upbringing, but his folks were as nice as could be and welcomed me in like I was one of their own. They barely spoke English, but I felt more at home with them than I did with my own family. I lay on their couch and slowly drifted off to sleep as the thought of those giant boobs danced around in my mind.

The next day, we headed out early and after he renewed his license, we drove to a train station where he told me he was going to take me into New York City. That, I thought, was awesome as it was the first time I had ever been to NYC and was eager to stroll around the crammed city streets and hail a cab on Park Avenue.

We took a train into NYC and ended up at Penn Station and then spent the whole day walking around the city: Time Square, ground zero, various shops and yes... I hailed a cab. It was so much fun I didn't want the day to end! Before we left the city that evening, both Castillio and I talked about the prospect of not returning to Camp Lejeune and it was I that persuaded him to come back with me, despite how much I hated it there. So, we decided to leave and arrived back at Penn Station at sundown and headed back to where his car was parked.

Once we reached the car, Castillio was shocked to realize

that he did not have his car keys! "Fuck! How are we going to get back?", he proclaimed angrily. I remember feeling defeated and almost started rethinking my decision to go back to base.

We ended up spending the next few hours trying to open his trunk to see if he had left them in there. I remember it was hot, and I took my shirt off to cool down as we continued trying to break into his car. At one point, I ran across the parking lot to grab a large stick to try and pry the trunk open with and on the way back, running full speed with stick in hand, Castillio mentioned to me that I looked like the "real deal". When I asked him what he meant by that he further explained to me that he thought I looked like a real Marine. Maybe I was a real Marine?

Eventually Castillo had to call a locksmith to make a key so that we could get back into his car. $250 later and we realized, once we opened the trunk with the new key, that the trunk was empty and it was clear that Castillio had lost his keys in NYC somewhere. He took the loss, and we drove out of the parking lot and ended up going to another strip club, which was essentially the same experience as the night before – rather uneventful.

Next, we drove to a pizza place which was situated in a very shitty part of New Jersey. We ate at the place and then decided that it was time to go back to his house so he could grab a few things before heading back to North Carolina. On the way back to his house, Castillio began to act strange and when I questioned him why he was acting weird, he told me that the pizza place we ate at was known to put hallucinogenic drugs in the food. I thought it was nonsense, but shortly after I also started to feel weird myself… was it in my head or was it real? I was hearing growling noises, loud yelling, and seeing weird things in the shadows as I sat in his car waiting for him to come back out of his house.

Once he came out of the house and back into the car, I noticed he was sweating and that his eyes were dilated. "Dude, your pupils are fucking huge!" I proclaimed. "Yea, so are yours!"

Fuck, we had been doused!

We were both tripping on some type of drug and ended up passing out on the side of the street for a solid 6 hours. Once we woke up, we had a laugh about it and drove back to the base with no issues.

The next weekend, Castillio asked me if I wanted to go back with him again and it was a hell yes from me! We began to drive and made it halfway through Virginia when his car began to act weird, making a loud rattling noise from the engine.

"I think we better turn around!" he said, catching me off guard. The last thing I wanted to do was go back to that base to sit in the barracks and hate life. I didn't want to go back and stew in my mind on all the ways I could slowly torture and kill Corporal Richards.

"Don't be a pussy!" I said, "we will be alright! Let's keep going!" He continued to drive, reluctantly, until the engine finally blew a hundred miles or so outside of Richmond. We were toast! There we were, sitting on the side of the highway at 8pm with no one to help.

I remember feeling bad because I had pushed him to keep going when he really didn't want to. I apologized to him, and he told me it was okay because he had also wanted to go and wanted off that base as much as I did. By some stroke of luck, Castillio had a distant cousin who lived somewhere near Richmond and even though he was not that close to him or his family, he made a call, and they were on the way to pick us up.

It was his distant aunt who came driving out to pick us up on the side of the road and she drove us to their house which was situated in an affluent suburb of Richmond. They barely spoke English, and I always wondered how they were able to afford such nice accommodations when they could barely speak the language, but I didn't really care at the time and was just glad to be off the base; I didn't really care where I was or what I was doing or who I was even with.

I slept in a nice bed that night, and the next day was spent in the house hanging out with his cousin where we shot bb guns at bottle targets in the back yard. Later that night, his cousin, who was a less than upstanding citizen and had in fact recently been released from jail, took both Castillio and me out to a bar. The place he took us to was one of those 21-and-older bars and the only reason I was able to get in was because his cousin had a hook up there with one of the bouncers. They let all three of us inside with no hesitation and once inside, no questions were asked about our age, and we were able to drink as freely as we liked.

It was a Hispanic bar, and I found out quite quickly that I was the only white guy in the joint. I was certainly out of place, but somehow, I didn't feel worried or threatened in the slightest. Castillio's cousin disappeared quickly after our entry to the bar, and we did not see him for the rest of the night. Castillio spent most of the night attempting to get some cute girl in a red dress, who he simply referred to as "Red", to sleep with him while I sat, looking around and drinking weird Mexican beer.

The bar was full of people and that included many beautiful women. I remember looking around trying to find a woman that I could potentially talk to, but it appeared as if though I would be out of luck. Since I was the only white guy in the place, I figured not too many of these women wanted to get with a "gringo", and so I just sat drinking beer and watching Castillio attempt to woo "Red".

As the night went on, I noticed that one of the bar ladies, a pretty waitress in a cocktail dress, kept eyeing me and making passes at me every time she walked near. She was relatively pretty with long, dark and wavy hair but she was certainly older than me by at least ten years. She smiled and winked at me when she made another pass, and I figured that it was my only shot to get with a woman in that bar. Towards the end of the night, Castillio had struck out with "Red", and the mysterious cousin had made his way back to us after finishing whatever shady activities he was involved in that night.

"Fuck, no luck here!" both Castillio and his cousin stated.

I took the moment to flaunt myself a bit and leaned into Castillio proclaiming, "Watch this. I'm gonna get that cocktail waitress!" They both looked at me, then looked at each other and laughed. Little did they know, I had watched the cocktail waitress make googly-eyes at me all night, so I knew that all I had to do was smile or wink at her and she was mine.

When she rounded the corner and headed towards me, I worked a little magic with my eyes and smile and that was all it took! She was indeed mine! She set her tray of drinks down, approached me, and rather aggressively grabbed me and pulled me into her – our faces practically rubbing against each other.

She was kissing my neck right there in the bar with both Castillio and his cousin in the shadows just watching it all unfold.

"Hola!", She whispered sensually.

It was clear she spoke no English, and I didn't speak any Spanish, so I just went with my gut and winged it.

"Eyyyy!" I said, slickly.

"Como Estas?"

"I'm good!", I said as smug as I could, feeling like James Bond himself. She then grabbed me and began to make out with me, and I could hear both Castillio and his cousin snickering at the display. Either they couldn't believe it or they were disturbed by the sight.

I swear she was about to pull me into a broom closet or something when out of nowhere, the bar was shut down and people began running all over the place.

"The Cops!", some guy screamed, "everyone get out of here!"

In the chaos, I lost my cocktail waitress and found myself back outside with Castillio and his cousin. It turns out that some poor sap got his ear licked off with a beer bottle in the parking lot and that was enough to bring cops out and shut the place down. Damn, so close!

Once we got out of the bar and into the parking lot, Castillio's cousin got into an altercation with a "rival" gang which were standing near his car. There were several other Hispanic men throwing insults in Spanish, and even though I couldn't understand what they were saying, I could ascertain that they weren't trying to be friends. Castillio readied himself for a fight and told me to line up next to him and his cousin.

I really felt like it was not a good idea for more than one reason. First off, they could have guns, knives, or a number of other weapons. Second, I did not feel like getting arrested for a cause that I had no stake in and potentially throwing my life away or getting in trouble with the Marine Corps for some stupid bullshit. "I don't think we should, dude", I said when he attempted to get me to join the both of them. "Look, you'll be family to us here if you join in. We have to – I have to. I don't have a choice!"

I thought about it and brought up my concerns to Castillo by telling him that because I was the only white guy involved, I would most likely be a major target and if something *did* pop off, I would be the first to be greased. Castillio pondered for a moment and ultimately agreed. Even he knew that me fighting with a random Hispanic gang was not a good idea in the slightest. Ultimately, nothing happened, and the rival gang left the parking lot, and we drove back to his cousin's house.

Back at their house at around 4 o clock in the morning, Castillio was talking to his cousin about the Marine Corps' way of life and mentioned that I had a five-year infantry contract. "That's two deployments for him!"

It was true! Five years was a long time and basically guaranteed I would go on at least two deployments. Most everyone else had a four-year contract and with luck those Marines could be out in two and half years with all the programs that allowed short-time Marines to get out early.

Five years, I thought dismally. *How will I ever get through it all?*

At the end of the weekend, Castillio's aunt drove us back to base after which we both departed back to our respective barrack rooms. It had been a great couple of weekends, and I was grateful to Castillo for taking me along with him, but I knew that it would also most likely be the last good weekend that I would be having for a while. As it turned out, it was indeed the last time I had any fun during my time in the Marines.

CHAPTER 7

ALL WORK AND NO PLAY

HUMPING AND FIELD OPS

Hot! I remember that it was so fucking hot! I was now doing what an infantry rifleman does in the field. There I was - a new boot in an infantry unit out in the field with guys that were way more experienced. I knew they would not take it easy on me and in fact, would do everything in their power to make my life as much of a living hell as they possibly could. God was it hot! I'll never forget that.

The first time going out, the sun beat down on us all with the strength of hell itself. Full combat gear - at least 80lbs, and miles and miles of humping. We would be humping, as a platoon, into a planned-out area where we would ultimately set up CP and then bivouac. Being the boot, I was subject to not only carrying all my gear, but the extra shit nobody else wanted to carry. I carried extra ammo, water, and two AT-4s. Each platoon took two AT-4s which were single use bazookas. They only weighed 14 pounds, which doesn't sound too bad, but because I had been ordered to carry both, it would tack on an additional 30lbs onto my already cumbersome load. I ended up slinging them crisscross over my back and then slung my rifle around my front. It was definitely not the most comfortable way to carry all the weight, and I could already tell the hump was going to be miserable.

As we took off on the hump, Corporal Richards would not stop laying into me with his taunts and insults and he continuously screamed and yelled at me the whole way. Once again, I could not move fast enough for the guy even though I kept up with everyone else. They were humping at a ridiculously fast pace, though, and it became apparent after a couple miles of humping that I was going to eventually struggle to keep the pace. After a couple hours, I began to fall out a bit mainly because I was struggling to breathe properly due to not only the incessant heat, but the two rocket launchers slung around my back which were choking me. Eventually, Sergeant Walker got onto my ass and started screaming at me, boot camp style, to keep up. He was also making me scream, "Aye, Sergeant!", every time he gave me an order.

We finally stopped for a rest and that's when Corporal Jacobs, my squad leader, decided to take the two AT-4s off my back and carry them the rest of the hump.

"You're slowing us down, Verlice" he barked.

I was incredibly grateful that he took those AT-4s because had he not, I may not have made the rest of the hump. It amazed me how they actually expected me to be able to carry all that weight on that long of a hump. Nobody else was doing it, but I guess it didnt matter. I was the boot and that was all there was to it.

Without the cumbersome weight of the AT-4s, I was much more nimble on my feet and able to move a lot quicker. Within a few hours, we reached our destination where we would bivouac, set up CP, and push patrols out and so forth.

I remember once we got to the destination we were allowed to take our packs off and sit on them while the Corpsman for my platoon, Doc Rudy, went around checking all of our feet and making sure that none of us were dehydrated. Corporal Jacobs approached me and chuckled stating, "Damn, Verlice. Those AT-4s killed me too!" So, it wasn't just me – even an experienced combat veteran could be broken off.

Around that same time as we were all sitting around doing nothing, Corporal Richards approached me and begun to fuck with me. He made fun of me, belittled and berated me, and physically tormented me with various physical maneuvers and other gratuitous activities.

I remember him getting so personal with his attacks that I was feeling terrible about myself. "You're a fucking piece of shit, Verlice. You can't even keep up in the hump!", he would say as another senior Lance Corporal would throw dirt in my face.

After the taunting and belittling, I was then ordered to dig various holes and set up some other things in the bivouac area until the sun eventually began to go down. I remember being so hot and miserable and wanting only to jump into a huge pool of icy cold water. I kept drifting back to summers of my childhood when I would go to the springs with my mother and sisters. The water would be so cold that I could only stand being in it for several minutes before slowly making my out. I remember wanting to go home so badly so I could jump into that icy cold water with my sisters again.

When it was finally time to sleep I was given shifts to guard various posts around the area including perimeters and other things of interest. This was pretty common for a boot in the field and a new Marine could expect to be put on every shitty watch nobody else wanted to do. On that night I remember I had to guard a large stock of ammo that was positioned in the center of the area we had cleared for our operations.

It was sort of like "fire watch" in bootcamp only much longer and more difficult. Out there, exposed to the elements, exhaustion, and total body pain, it made bootcamp fire watch seem like a walk in the park. I had to patrol around tired as hell with my rifle and full combat gear until it was the next Marine's Shift. It felt like I was patrolling around forever and I remember checking my watch *-0300, Damn! Where the fuck is my relief?*

All I wanted to do was sleep and try to recover from the grueling hump that left my body fatigued and sore like I had not felt before. Finally, another boot Marine came to relieve me, and after I briefed him on everything he needed to know, I made my way back to the hole where I would be sleeping. I threw off my flak and Kevlar and slowly laid down in my bivy sack and drifted off into a very unsatisfying sleep, my rifle still slung around my front.

SO HELP ME, GOD

I remember only about an hour or so later, I was awakened by the screaming of my company Master Sergeant, Master Sergeant Sneed. Master Sergeant Sneed was a very intimidating Marine and had in fact been an ex-Drill Instructor, meaning he had that distinctive scream that I was all too familiar with. Apparently, the last Marine who had relieved me on guard did not wait for another Marine to relieve him and instead just went back to his hole while the ammo dump was completely unguarded.

A Marine NEVER leaves his post (fifth general order: to leave my post only when properly relieved), and doing so was punishable by potential NJP (non-judicial punishment – or article 15 as it's known in the Army). Master Sergeant Sneed was screaming like a madman over the guard mishap, and I can remember sinking into my hole, freaking out internally that I was somehow going to be in trouble since the Marine being chewed out had relieved me. I was worried that in desperation he may tell the Master Sergeant that it had all been my fault that the ammo dump was left unsecure.

Luckily that did not happen and it was not my issue as I had made sure my relief had properly taken over the guard shift before going back to my hole and I was not called out for it. The boot Marine responsible was punished and spent the rest of the night guarding that ammo dump alone. I remember the Marine's name but I will refer to him as Booker. Booker was a weird character and seemed very meek and mild, green, and scared.

It didn't take long living in the field to realize the toll it eventually takes on the body. We punched out patrols on a daily basis, did perimeter watches, dug in various positions and so on every single second of the day. The heat would be so bad that everyone would be laden with prickly heat rash and other maladies such as athletes foot, shin splints, and crotch rot. About two weeks was all it took for me to ditch the underwear and start going "commando" due to the intense chafing experienced on the nether regions. At one point, my balls had gotten so raw that I literally had to peel them off the side of my leg which took skin and all with it. We never took showers when we were out as we simply were not able to and the best we could do was take a "grunt" bath with baby wipes.

For me, that period of time was when I started to see more of the comradery the Marine Corps always boast about. We really had to rely on each other for warmth when it would get cold and motivation to keep going when it was scalding hot and I quickly found out that nothing is considered "gay" when in the field. There were times when we would be wet, tired, and cold and had to cuddle up with other Marines just to keep body warmth. I remember times huddled in a circle while we all passed around a single cigarette and told jokes to make each other laugh. Some of those moments were very trying for us all and we only had each other to lean and rely on.

Another thing I realized quickly during that time was that the Marine Corps didn't tell us shit. We never knew exactly where we were going or what exactly we would be doing and every time something would be "sprung" on us last moment, we would get increasingly more pissed off. There was always "bumscoop" going around and some of the guys would say, "oh, I heard they're going to make us go here," or "I heard they're going to make us do that." Most often it was nothing more than bullshit rumors, but what was true was that we almost never knew what was next.

Sometimes we would be told transport would pick us up, or the Osprey aircraft were going to rendezvous with us at a certain point only to find out that we had to hump another 10 miles to a different bivouac area or perform some other patrol in a new area. It was frustrating for us all but there was nothing we could do except deal with it and keep pushing on – there was no quitting as an infantry Marine. It was also during some of those early days in the field that I was constantly put on what is called "point".

Point, walking point, or being point man, is the position of the first man in the squad formation of a patrol. The point man leads the patrol, sets the pace, and must ensure that they are keeping a keen eye out for any potential hazards such as enemy, IEDs or booby traps, and anything else which could be a potential issue to the patrol. Obviously, as a boot I did not have the first clue how to walk point, and it was pretty obvious. That didn't stop them from making me do it all the time mainly because they just wanted a reason to mess with me or point out my flaws. I hated walking point because I didn't know what I was supposed to be doing and luckily after a while, they stopped making me point man and then I was usually the last guy on the patrol – bringing up the rear, as they would say.

Every night would be the same with constant rotations of perimeter watch, ammo or chow watch, and LPOP (listening post/observation post) watches. I would always have the shittiest watch too, obviously because I was the boot, and it was usually some middle of the night 2–3-hour watch on the perimeter some 100 yards or so away from everyone else. Perimeter watches would be down in a hole or some other established position, sometimes alone and sometimes with another Marine, and basically consisted of simply sitting there with the gun out locked and loaded in case of any attack or ambush. I remember it was always somewhat scary because it was so dark, and I always wanted to go to sleep while on it – a major hazard for more than one reason. Sometimes, I would have my whole poncho liner down with me on the perimeter where I would drape it over my helmet and shut my eyes, dozing off while staring down the sight of my rifle. That was a huge risk I was taking but was unavoidable due to the extreme exhaustion and lack of sleep overall. I simply could not keep my eyes open.

The annoyance of being a boot really began to kick in after several patrols and stays in the field. For example, being on perimeter watch, which we would call either left LPOP or right LPOP, I would have to report via radio any movement, sights, or any other goings on in the area. Every time I would get on the radio, I would need to report "PFC Verlice reporting from left LPOP…movement in tree line" or whatever else was needed to be reported. There would inevitably be some senior Lance Corporal or Sergeant on the other end who would proceed to mock me and make fun of my voice or any number of other trivial things which they thought was funny.

That type of stuff got old quick, and it was just another thing that could really get under my skin when it was already miserable to begin with. Those things, as insignificant as they seem, were just a few of the things that put me at odds with my decision to be a United States Infantry Marine.

CORPORAL RICHARDS

One night when the whole platoon was humping to another location, my platoon had two Marines go down. It was a long hump through the middle of the night in difficult terrain after two weeks out. One Marine went down as a heat case – meaning he had heat stroke, and another received an injury to his foot sufficient for a medical transportation. The Marine who received the foot injury was my fire team leader, Corporal Richards. Even though he was injured, I remember thinking, *HA, all that fucking around with me and look at you now! You can't even walk!"* It honestly felt like justice given the hell he had given me earlier on my first hump out with the platoon.

Despite the cruel person Corporal Richards had been to me earlier and despite my dislike of him, I was the one who went back and helped pick him up and hobble him over to where he needed to go. I'll never forget his arm around my shoulder as he helplessly leaned most of his weight on my own body, it was like he had no fight left in him. When I got him to where he needed to go to be transported for medical attention, he looked at me and said, "Verlice, I'm sorry. You are a good kid!" He paused for a moment and continued, "You're a good Marine."Him saying that meant the world to me and any anger or indignant feelings I had towards him at that point melted away and I felt nothing but love for my fire team leader.

Corporal Richards never really recovered from his injury which included his foot and whole leg, and he never led me as a fire team leader again. I never really saw him much after that and he was ultimately medically discharged sometime later for these injuries as well as some other injuries he sustained from an IED blast. Unfortunately, I found out several years after I was out of the Marine Corps that Corporal Richards committed suicide due to the long-lasting pain he endured, PTSD, and lack of care from the VA. I pray Corporal Richards found peace and is resting in heaven.

After that incident, my fire team was assigned a new fire team leader, Corporal Duggins. Corporal Duggins was a laid-back guy and was about 9 months shy of being done with his enlistment in the Marine Corps. I was still low man on the totem pole; however, things were beginning to get a bit easier for me in terms of the "boot" treatment. Just like in bootcamp, I began to get into a flow where I sort of found my groove and just co-existed – chow to chow as they would say. It was still physically demanding though, and that never got any easier.

I remember one day when my fire team was tasked with MOUT (military operations on urbanized terrain) patrolling, I *nearly* went down as a heat casualty. It was one of the hottest days I think I have ever experienced, and the weight of all my gear made it difficult for me to catch my breath, not to mention my fire team was zipping through the place like madmen with not a single second to break. I was hydrating; however, it was not enough, and I could feel myself slipping further into the danger zone. I knew the early signs of heat exhaustion: nausea, rapid heartbeat, profuse sweating, and I had them all. Even though I felt as if I was going to go down, it did not stop me from continuing with my fire team's mission.

When the whole platoon was tasked with MOUT, there would be different fire teams in all different directions of the urban area. My fire team was tasked with patrolling roads and clearing various compounds which included breaching doors and clearing certain areas including rooms and hallways. Eventually, I started to feel as though I was going to pass out but due to fear of being reprimanded, I decided not to say anything and pressed on. If it had not been for Corporal Duggins looking behind him and seeing my purple face, I would have fallen in just a matter of minutes. He immediately called for a Corpsman, and I was taken to a shaded area where my gear and flak was removed. The second I was free from the cumbersome gear, I felt 50% better and could start taking deep breaths to slow my rapid heart rate. I could feel a cool breeze hit my soaked skivvy shirt and nothing had ever felt better.

Doc Rudy came over along with another Corpsman who both determined that I was dehydrated but not in critical condition. They administered an IV of lactated ringer and made me consume a weird hydrating liquid which tasted horrible but worked like a charm because I pissed nothing but pure water for the next four days. I remember Doc Rudy looked at me with so much anger and dumped a full canteen of water over my head (which is not something they were supposed to do to heat casualties).

"Fucking drink water, Verlice. What do we always say!".

I looked at him, feeling mortified, "Aye, Doc!".

As I was sitting there getting treatment, I remember seeing First Sergeant Kellie scoping out the area. He had been moving between patrols and fire teams, which was uncommon but sometimes did happen, and he stopped when he saw me crumpled over in the shade. I remember him looking at me; he looked at me with an almost impish grin.

Why the hell is he looking at me like that?

He didn't say anything and eventually moved on after staring at me for a little while longer. I remember thinking it was strange but was also happy that he didn't say anything to me.

After I recovered enough, I joined my fire team again and we rested near a tree and ate MREs. I remember receiving a lot of shit about the heat incident and for the next couple weeks, it was all I heard about every time I spoke to a senior Lance Corporal. Incidents such as that, as well as others, spread around the ranks like wildfires and those would often lead to vicious rumors among the senior Lance Corporals and boots alike. It was one reason why I didn't attempt to disclose anything that happened to me later.

Everything the senior Lance Corporals taught us about combat and combat situations was in stark comparison to what the combat instructors at ITB had taught us. I remember the senior Lance Corporals would always say, "throw all that shit you learned in ITB away because *we* are going to show you the *real* way to do this shit!" Even though I was just a 19-year-old boot, I still had enough brains and common sense to know that the things they were showing us were not only incorrect and wrong, but also dangerous and counterintuitive as well.

For example, when clearing rooms and specifically hallways, the senior Lance Corporals had a tactic which they called "*shoot the boot*". The idea was that when a Marine fire team needed to go around a corner in a building complex and it was unknown what was around that corner - be it an enemy with a PKM or any other hazard - the most junior Marine (boot) would be the first to "*shoot*" across the hallway thus getting the potential enemy's attention. The point of that move was to "distract" the enemy so that the next Marine behind the boot can "*pop*" out from around the corner and shoot the enemy before the enemy can fire a shot at the boot. The meaning of "*shoot the boot*" was two-fold as it basically meant the boot would be the one to be shot if in fact there was Hodji with a PKM around the corner. The senior Lance Corporal would be damned if they'd risk their own life like that...thus "shoot the boot!".

Another thing that was morally wrong, in my opinion, was the method in which detainees were instructed to be taken. It was common for a Marine infantry patrol to encounter suspected Taliban and those that were suspected were often restrained and taken for questioning by some other unit. The way they taught us to do it was by ordering the boot to zip tie the hands of the detainee so tight that they would be likely to bleed. They wanted us to man handle them to the point of breaking bones or doing serious damage. They would say, "if that fucker moves even a centimeter, jam that rifle barrel so far down his throat that you blow the shit straight out of his ass." I remember not performing the task "aggressively enough" and then having it performed on me by one of the seniors for my own "learning purposes". Obviously having it done to me was unpleasant and from there on out, to avoid having it performed on me again, I would just be as aggressive and violent as possible.

They also taught us interesting ways of checking dead enemies and that was done by "sternum tapping" them. The seniors would instruct us to jam our rifle barrels so hard into the sternum of an enemy – to check for life - and if they were alive still, it would be incredibly obvious, and they wouldn't be alive for much longer. Those were some of the more gruesome things which the USMC infantry was involved in and were very commonplace during that time of the war. All those things never really bothered me at the time, but looking back in hindsight, it all seems macabre.

CONTACT

Everyone who hasn't served often thinks that war is like the video games, "Call of Duty". Hell, even I had those fantasies before I joined. What nobody ever considers is the fear, pain, exhaustion, tension, fog of war, missing home, missing a girlfriend or parents, being hungry and thirsty, the not knowing what is coming next. Taking contact is what happens when a Marine unit or patrol comes under fire from an enemy position. Typically, the unit is patrolling in some formation, moving slowly and scanning the environment for any movement, unusual patterns, or anything of suspicion. Everyone on the patrol turns around every few moments to check their backs as the unit slowly snakes its way through the terrain.

A Marine infantry patrol can take contact from the front, right, left, rear and where it occurs depends on how the unit reacts. When a Marine infantry unit comes under contact or fire, someone will scream, "contact front!" or "contact rear", or "contact left!", depending on where it comes from and immediately upon hearing the word, everyone in the unit will echo the words back and run towards the direction where contact came. As Marines, we ran towards the bullets to neutralize the threat instead of setting up perimeter security or performing some other defensive position like the Army typically does.

Walking in the rear of the patrol as per my usual, I heard, "CONTACT FRONT!" being screamed from up the line. In an instant, I knew what to do as its simply muscle memory by that point.

Everyone began running towards the front and into a huge field where the contact would have come from. Bullets were ringing all over the place and I could hear machine gunners from weapons company which were positioned on a different end of the field blazing away. Because I was at the rear, I had the longest way to run to get on line where everyone would start a sweeping motion to neutralize threats. Running through the field to where I needed to go, all I could think about was the pain and exhaustion I was feeling. Never would I have ever worried about being shot or killed; to the contrary, I felt death would have been welcomed had only it been swift. Pain always seemed to outweigh any other notion one may think they would have going through their head in a moment such as that.

When I finally got to where I needed to go "on line", I plopped myself onto the dirt and mud and laid in the prone trying to catch my breath so I would be more accurate with my shots. The Marine next to me was firing and I began at once to lay some rounds down range as well. What I was shooting at…who knows? It could have been a balloon tied to a post– perhaps it was to me. My mind went to a different place, and nothing felt scary or unreal; it was all second nature to me at that point and was nothing more than just "*training*" in my own imagination. Pure muscle memory and nothing more.

After laying some covering fire down we began to "buddy rush" to the target. Buddy rushing is essentially where one Marine will cover the Marine next to him while he runs a bit closer to the target. Once he gets down and begins covering fire, then it's the other Marine's turn to run and that continues all the way until the target is no longer a threat or the "cease fire" command is given. I covered fire while the Marine next to me ran for three seconds. Once he began to lay down cover, I got up and ran, saying in my head

I'm up…he sees me…I'm down!

That is how long I ran— only three to four seconds. The longer a Marine stays up, the more of a chance of being shot they have. As I ran, bullets whizzing all over the place, my mind went back to Parris Island when SSGT Stubb had me performing incoming fire maneuvers down the rifle range the morning of the chemical bath.

Bullets continued whizzing all over the place as we kept buddy rushing closer to a dirt road where the "target" was. Once we were in position close enough, we began to all lay down fire in the direction of the "target". I looked through the ACOG (scope) of my rifle and scanned the horizon and there it was… a *target!* I had to shoot it; I wanted to shoot it. I aimed and squeezed the trigger – miss! I shot once more and missed again. I was breathing heavily as pain crept up all over my body. It made me more unsteady and I tried to control my breathing long enough to get a good shot off.

Finally, I steadied myself and got up onto my knee which would have been considered dangerous with bullets ripping through the air, but it was the position I could be the steadiest and it gave me the best view from the tall grass I had been prone in. It gave me the best chance to hit a target. My head *would have* been a target, but I didn't fear at all – nor would I have ever. I looked through my scope and saw the "*target*" pop out.

FIRE!

MISS!

Fuck!, I thought as I quickly reloaded my weapon.

I steadied once more, aimed slowly and when I saw the "*target*" once more, I pulled the trigger.

HIT!

I saw it "*pop*" – clear as day. I threw myself back on the deck and breathed a sigh of relief, feeling good that I had done my job that day.

"CEASE FIRE!", I heard being screamed over the line.

It was over! Ten minutes is all it took. That is what the Marine Corps infantry is – all that training for ten minutes.

Afterwards, we "policed the area" which was the roadway where the "targets" had been, and where the contact originated. Nothing there at all! There were radio calls, some standing by, and then we moved on down the road where we continued with our objective to which I don't even remember, probably some chow or something. Soon we would be moving back to the rear where we could rest some.

It was then at that point that the gut feeling I had at both bootcamp and ITB graduation would make itself known and the hell that was waiting for me – destined for me – would manifest and set me on a downward spiral laced with harassment, abuse, and torture which up until that point in my life I had only ever read about or seen in movies. That moment right then moving forward began my war – my hell.

CHAPTER 8

A FORCEFUL CHANGE

"GEAR IN MY REAR?"

We were ordered to wake up early one very hot morning and little did I know, everything for me was about to change in ways I could never have imagined. The day proceeded with one of the most grueling death marches I had ever and will ever partake in. The march included several of the other boots and myself within the platoon carrying all the extra gear which was to be moved to the rear.

Some of the gear to be moved included water and oil gallon jugs as well as other heavy equipment which I thought would have been better left for transport. Bumscoop had gone around the ranks and of course we were all under the impression that the event was going to proceed differently, such as being provided transport; however, that turned out to be nothing more than bad word per the usual.

We were expected to hump everything from point A to point B at a superhuman pace, in extreme heat, and under the worst pressure imaginable. As the morning went on and the heat dialed up, tension began to rise in the platoon and soon, people began to suffer. I was physically doing okay but was quickly being pushed to my limits as we were soon being instructed to run with all of the equipment in order to make some type of good "time". Again, that type of thing was not typical but because the higher command felt they had something to prove with all their prior negative history, nothing else mattered – not the men, not the gear, nothing!

It was then that things began to take a turn for the worse and some Marines began to go down - mainly the boots. Somehow, I was still ambulatory but was rapidly declining as the run was almost never ending. As the event drug on, one Marine began to struggle more than anyone else and that Marine was Booker, the Marine who left his post at the ammo dump a couple months prior and had been chewed out by Master Sergeant Sneed. Booker was way behind everyone else with ammo cans which probably weighed 50-60 pounds, and he looked as if he would die.

I turned around and made my way towards Booker, the phrase no Marine left behind buzzing around in my head, and when I reached him, I could see how bad off he *really* was. He could no longer operate his legs and was white as a sheet. I grabbed him by the collar, and he lumped his weight onto me as we trudged our way forward bit by bit, trying our best to run as fast as we could. It was then that Booker began to uncontrollably and copiously vomit, and he didn't stop vomiting the whole time I was helping him.

It was the most macabre and grisly thing I had witnessed in my life and for some reason, out of everything else, stuck with me the most. The sound, smell, and sight of it all imprinted on my brain where it would make itself home for the rest of my life. He couldn't breathe, and I honestly thought that he was going to die. Despite my fear and waning motivation, I moved him along with me as best I could. Soon, his vomit turned to pure blood...

BLOOD!

It washed over my arms and dangled in dark crimson tendrils from my fingertips. The smell was like nothing I had experienced before – chemicals and sour lemon juice. I had Booker's bloody vomit all over my arm and immediately went into a panic attack, realizing that it was I who could no longer breathe.

Despite my own pain and panic, I continued to carry Booker along as he vomited blood without stopping even for a moment. I fought panic and began to slow down, unable to move faster. My heart was beating a mile-a-minute, and I thought at that moment I may collapse. I was in full blown panic mode, unbeknownst to me, and began to walk and then crawl with Booker who then collapsed onto the dirt. A corpsman was called up and handled Booker and then picked me up and began to run with me. That *freaked* me out more than anything because it put into my mind the notion that I may suffer the same fate as Booker.

I immediately screamed to the Corpsman, "PUT ME THE FUCK DOWN!", all at once leaping from the corpsman's arms and running like a man on fire. Somehow, I had a burst of adrenaline and soon found myself moving back with the rest of the Marines.

I was still in a state of panic when I was approached by Master Sergeant Sneed, the ex-drill instructor who had handled booker a couple months before. "FUCKING MOVE, VERLICE!", he screamed, viciously. "AYE, MASTER SERGEANT!", I screamed back, finding that I was having trouble screaming due to the panic attack which was still coursing through my body.

Why Master Sergeant Sneed began to latch onto me, I'm not sure as I had done nothing wrong and did only what I was told that day. As a matter of fact, I had helped Booker, so why he was screaming at me was mind boggling. At that point, I was gone – simply gone. What I saw from Booker was enough to shut me down permanently and I found that moving forward after that day, I was no longer able to run or operate the same as I had been prior. I lost myself in that moment – just lost all sense of who I was, where I was, and what I was even doing there.

Somehow, I made it on foot with everyone else to our destination "in the rear" and to the spot where they wanted all of us and everything staged. Booker was nowhere to be seen and I'm assuming they lifted him somewhere for treatment. All the Marines were "standing by" in a grassy area, some half dead and some in better shape than others.

I'll never forget that one tree and that sky…that sun - how hot it was that day, and how fast my heart was beating as the thought of everything I witnessed that morning flowed through my mind! I stood by that tree, catching my breath and coming down from the panic attack I had been overtaken by, and tried desperately to find some solace. I really did not know what I was going to do, but I knew that something was different and I had changed somehow. I believe I had reached my limit at that moment and was at the point of a total mental shutdown.

It was then that I was approached by Master Sergeant Sneed, who for no good reason singled me out over all the other Marines standing nearby. Sure, he had gratuitously latched himself onto me during the movement to the rear and even though I did everything I was told and made my way back with everyone else, he felt the need to go one step further, and for what reason?

Well, I still don't know to this day.

MASTER SERGEANT SNEED

As I stood there, trying to quietly calm down, with the relentless sun beating down on me, Master Sergeant Sneed approached me with a sour look of hatred and rage perched upon his face. He spared no time in getting straight to his point with me and proclaiming to me what he wanted. He stood in front of me with his hands on his hip in anger, almost the way a mother would do if you misbehaved or broke a rule which then warranted punishment.

My heart began leaping out of my chest as I figured he would either yell or berate me in some way - the way only a Master Sergeant could do. To be honest, it was the first time since I had been in the Marine Corps that I was approached by such a high-ranking Marine and since I knew what type of ass-chewing he was capable of, I was scared out of my mind. What he attempted was far worse than any ass chewing could have ever been.

"VERLICE!" he barked, getting my attention immediately. "Yes, Master Sergeant?", I responded back without haste.

"Pull down your trousers!"

I froze, not knowing what to say or think. *Am I hearing him correctly?* I thought in a daze. *Did he really just order me to pull my pants down?*

I stood there frozen in a mix of incredulousness and terror as I wracked my brain trying to make sense of what he had just ordered me to do. "I said pull your *fucking* trousers down NOW... skivvies too... and bend over!"

He has to be fucking joking?

My heart was thumping out of my chest as I looked around frantically, hoping someone would hear what was going on and try to step in to stop it. Nobody saw nor did they care and even though Marines were within ear shot, not as single one gave any attention. Nobody would have dared questioned the command of a Master Sergeant, especially one as mean as Master Sergeant Sneed.

In his hand he held an object; it was long, thick, and made of metal. I believe it to have been some sort of medical device, like a metal speculum or something of that nature. In laymen terms... it was an asshole spreader.

"This is going up there!" he demanded, "and this Corpsman will watch!"

There was a Corpsman next to us, and he looked at me with the same fear that I probably had in my own eyes. It was from the Corpsman that the dreaded device was procured – from his little bag of goodies. The Corpsman didn't say anything, but his eyes conveyed the message, *What the fuck did he just say?*

Master Sergeant Sneed looked at the Corpsman and said sternly, "RIGHT?"The Master Sergeant had clearly thought it all out to some extent because the act that was going to be committed was going to occur in *front* of the *whole company*. Clearly, in order to cover his bases and ensure that nobody could accuse him of doing anything illegal, he had the Corpsman monitor the situation so that it could look as if though it was a "medical" examination of some sort.

The Master Sergeant was *not* a doctor, nor did he have any reason to have such a procedure done on me as I was clearly not injured, unconscious, or even outwardly in distress in any way that would warrant my ass being spread. The order was simply gratuitous in every way.

The Corpsman hesitated and then let out a long sigh, obviously not willing to question the Master Sergeant's command lest his ass be the one spread apart next. I was begging for the Corpsman in my mind and with my eyes to say something, but instead he just shrugged and gave me a look which conveyed submission; sort of like he was thinking, *okay, kid…this sucks but we must get through this.*

At the time, I had no idea why he was trying to have that done to me because like I said, there was no reason for it. Looking back, I now must believe that it was either for some sexual desire or to assert some type of dominance – either way it was considered illegal and certainly constituted sexual harassment and assault.

As I stood there with the sun in my face and beads of sweat

dribbling down my brow, I was taken back to Parris Island, standing around the Iwo Jima Monument with my right hand raised as I repeated back the oath to my country.

...that I will obey the orders of the President of the United States and the orders of the officers appointed over me...

Surely that activity would not have been something the President would have approved of. Those couldn't be the orders the oath to my country were referring to, *could they*? Where is one supposed to draw the line? He *was* my superior and he *was* giving me a direct order... but the order was...sick!

What was I supposed to do?

Those were all the thoughts that zipped through my head as I stood there incredulously as Master Sergeant Sneed waited for me to comply with his demand.

There are a few moments in one's life when it's truly "do or die". There I was, a low-ranking boot being ordered to do something terrible and invasive by a Master Sergeant in the Marine Corps. All the training in bootcamp, ITB, and in the fleet as well as the incessant brainwashing which occurs in the Marine Corps told me that a Master Sergeant's orders are *never* to be disobeyed. I had never personally seen it done; I had never seen a boot (or anyone for that matter) say no to a Master Sergeant or any high ranker for that matter. I had two choices: bend over and spread them wide or say no and deal with whatever consequence would befall me. I was truly damned if I did and damned if I didn't. "I'm not fucking asking you! I'm telling you, Verlice!", he barked with an increasingly indignant boom.

All at once, everything I had done in my life up to that point be it with school, family, friends, or by myself became totally irrelevant as I felt my soul quickly being snatched from my body by an unseen entity. I could practially see my USMC career going straight down the toilet as I stood face to face with Master Sergeant Sneed - my executioner.

I quickly made up my mind that I would rather be killed right then and there than to have that guy shove anything up my ass. The pain and humiliation that something like that would cause me as well as the rumors and abuse I would receive from all the other Marines standing by would have been too much to live through. The guy wanted to sexually assault me in the most deplorable and ignominious way possible in front of the entire Alpha Company, most of whom would have had a front row seat to my "examination".

I dug my heals in and made up my mind. It was time to die!

"No, Master Sergeant. I won't!"

The look on his face conveyed disbelief. He had probably never heard a boot tell him no before, and he was instantly fired up, ready to chew my head off in an instant.

"NOW!" he screamed in a very distinct and familiar drill instructor-type voice.

Wow…the guy actually was a DI at one point, I guess.

Everyone in the area snapped their attention in our direction and then all eyes were on me.

The way rank worked in the Marine Corps was simple for

a boot: the higher the rank, the tighter one's asshole would get. There really was no leeway when given an order by a higher rank and a Marine, especially a boot, was expected to do whatever a higher-ranking Marine ordered of them. The problem I faced, however, was not so simple.

I again said as calmly and politely as possible, "No, Master Sergeant. I can't!"

Ain't fuckin' doing it, asshole!

He ordered me about five more times with increasing levels of hostility and each time I said, "no, Master Sergeant.", he got louder and louder.

Finally, realizing that he had lost the battle, he approached me and grabbed me by the neck. I thought it was all over at that moment, but he just shook me and screamed at me. He then dragged me to an isolated CP area – a building – where nobody was.

My fear rose to new levels as I had just defied the command of a Master Sergeant, which was unheard of, and had also been ordered to do something awful, probably also uncommon, and now I was going to be alone with the guy in a building where nobody could see anything happen. Anyone on the street or back home would say, "break his jaw!", but it didn't work that way in the Corps. If I would have done anything physical with the guy, I would have just been court martialed and put in the brig.

Once he got me into one of the empty rooms, he took a seat, and I was shocked to see that his whole demeanor changed. At the time, I really didn't think much about it due to shock and the day's events, but looking back it was clear that he was afraid. He obviously just sexually harassed me and attempted to assault me sexually and I believe he was afraid I would tell someone about it.

"What is the matter, Verlice?" he said, sounding sincere, polite, and concerned. Since I had assumed that he was going to do whatever he wanted to me in that building, his sincerity took me off guard and because of everything that had happened with the day – with Booker and then the harassment – I let the waterworks loose and began to cry copiously.

He looked shocked at my response to his question and acted almost as if he could not believe that I was crying. He had that sort of attitude that an older brother gets when they hurt their younger brother playing and they start to fear that they will get in trouble by their parents.

"What is the matter with you?"

He was hushed in his tone, like he didn't want anyone to hear what we were talking about. He offered me a banana and ordered me to eat it, "you're weak, Verlice.", he told me as I took it from him and began force it down.

Somehow that banana was supposed to make up for what he had just attempted to make me do moments ago. "Talk to me, Verlice," His voice sounded sincere and concerned, a stark contrast from the monster who had just ordered me to bend over to have something inserted into my rectum not moments prior.

It was then that I said to myself, *Fuck it all*

There I was, just 19 years old, going through shit that nobody should ever have to endure. I had volunteered to do what I was doing, and I had already put up with enough abuse and torment to last a lifetime, but I did not sign up for *that* - I did not sign up for sexual harassment. That was taking things way too far.

It was at that very moment that I made up my mind to leave the Marine Corps behind. I no longer felt the values or motivation to continue my life as a grunt in the USMC. The problem, though, was that one does not simply leave the Marine Corps like one does a regular job on the streets. It didn't work that way and I knew it, but I didn't really care at that point. I knew one thing was for certain: my time in the Marines was going to be over one way or another.

Even though I wanted out after the dramatic chain of events, I knew that I had to be careful because the Marine Corps loved to screw Marines, especially young Marines, and would do anything in their power to ensure that I would leave the Marine Corps with nothing and, if things went really sour, a bad discharge as well. I knew they would do everything to ensure that a black mark was left on me including diagnosing me with all types of mental issues or making me out to be a troublemaker so they could kick me out with a bad conduct discharge.

I immediately thought about telling the Master Sergeant that I was going to kill myself primarily because I knew of a Marine who had said that and it was game over for him – immediately sent to the hospital, left alone, and then discharged honorably. I also thought against it because I also didn't want to end up in a padded room or to be discharged on some negative psychological grounds.

It was a real sticky situation I found myself in.

When he pressed for me to again to tell him what was wrong, I simply squeaked out, "I don't want to do this anymore, Master Sergeant." He looked puzzled and questioned, "What do you mean? What don't you want to do anymore?"

"I don't know, Master Sergeant?" Here I was still forcing to address this freak by his rank and literally had nothing to say to him in way of defense. To him, and anyone else in the Marines, I had simply defied an order even though I was within my rights to do so. He questioned me, gingerly, a bit more and sort of came to the conclusion - on his own - that I wanted out of the infantry and out of the Marines.

The truth of the matter was that had it not been for that bastard trying to sexually assault me, I would have gone on my merry way even though the events of the day had been, what some may say, traumatic. I would have put it all behind me and carried on, but I drew the line at the sexual harassment. That is NOT what I, nor anyone else, signed up for.

He seemed to be playing mental mind games with me and as a result, I started to get a bit desperate and in my desperation, I blurted out, "I request to go to a hospital." I was really trying to avoid saying something that was going to leave a mark on me, but I was also trying to get my point across that I was not going to be putting up with what he had attempted with me earlier. It may have been my only chance to report what had actually happened.

The Master Sergeant simply looked at me and huffed, "What are you saying to me?" He was trying to get me to say that I wanted to kill myself. I just shook my head, "I don't know, Master Sergeant." I wasn't going to take his bate and quite honestly I did not feel like I deserved to be ushered into a padded room somewhere just becasue he wanted to play doctor with my asshole.

"You know what this means for you, right?", he grumbled, shaking his head. I didn't say anything to him and attempted to compose myself the best I could. He eventually got up without saying anything else and left the room. Where he went, I'm not sure, but I do know that he is the direct reason why the next chain of events occurred.

FIRST SERGEANT KELLIE

After several moments of sitting alone in that room, the silence was shattered by the back door ripping open. I sprung to my feet as I heard my name yelled out ferociously, "VERLIIIIICCCEEE!"

It sounded like the devil himself screaming out to me straight from Hell. It was First Sergeant Kellie who up until that point had not known my name and had never said two words to me. It was clear that Master Sergeant Sneed had briefed him on the situation to some extent.

Certainly, he had not told him the *truth* - that he attempted to sodomize me moments earlier with a speculum, or if he had, First Sergeant Kellie clearly didn't give a shit. The Marine Corps way of life was very similar to what I imagine prison life is like. Nobody would snitch on each other, and they would always back each other up no matter how fucked the situation was. I imagine Master Sergeant Sneed simply told First Sergeant Kellie that I had disobeyed a direct order from him and was now trying to "quit" the Marine Corps. Whatever it was that Master Sergeant Sneed told the First Sergeant made him mad enough to take every bit of pent up anger he had out on me with no mercy.

I approached First Sergeant Kellie, and he immediately grabbed me by the collar and threw me into the wall several times. He was screaming all types of obscenities at me - calling me a faggot, pussy, coward, traitor, and many other obscenities. I was totally and completely unable to do anything to get out of his grasp, defend myself, or even explain myself.

"What the fuck is your Goddamned problem, fucker?" he barked. Every time I attempted to answer him, he would just throw me into the wall harder. I should have screamed back to him, *"THAT FUCKER TRIED TO RAPE ME ANALLY, SO FUCK YOU AND FUCK HIM!"* But of course I did not; it probably wouldn't have made much of a difference anyway because the guy was on a mission and his mission was simple: Kill PFC Verlice!

"Mother fucker, go over to the wounded warrior battalion and see all of those mangled Marines and then tell me that you have problems!" I could not get a word out or even report to him what had happened and why I was there in the first place. I saw nothing but fire in the guy's eyes and knew that there was no fighting anythign which was happening to me.

The First Sergeant outranked the Master Sergeant and there was certainly no way I was going to defy both in one day. After the beating, I was released and headed back out to my platoon where, shaken, weakened, and beaten, I returned to my squad and tried to pretend that none of what had occurred happened.

To my surprise, nobody seemed to notice that I had been gone and furthermore, nobody asked any questions. I vowed to myself to never speak to anyone about what Master Sergeant Sneed had tried to force me to do, mainly out of shame and embarrassment, but also because I was in denial about the whole thing. That could *never* happen to me!

Even though nobody knew what happened, I noticed that everyone appeared to look at me differently. It was like they *did* know somehow but I'm pretty sure none of them actually did. I felt eyes staring at me; I felt everyone watching me. Everyone ate a big meal in the garrison that morning, and as I sat and tried to eat with Drummer and a couple other Marines, I spoke of nothing which had occurred that morning.

The thought of Booker's violent and bloody vomiting, Master Sergeant Sneed's awful command, and the beating by First Sergeant Kellie filled my mind, tormenting me with its violent intrusiveness.

What am I going to do now?, I thought dismally as I pushed nasty food around on my tray.

FROM THIRD TO SECOND

At some point after the event, I somehow found myself speaking with the unit Chaplain. It was something that I had never thought about doing and had not even known of any other Marine doing so either as it was sort of tabu. I really regret not being forthright and telling him what Master Sergeant Snead had tried to do but was in too much denial that it never even crossed my mind as I sat there in his office talking to him. The unit Chaplain was essentially a Marine officer but instead of two shiny rank insignias on his collar, one side bore a gold cross.

The chaplains were barely Marines and were usually civilians who received a direct commission right off the street for their valued skills. Those types of direct commissions were usually given to chaplains, doctors, and other highly sought-after professions needed in the Marine Corps. The guy had probably been through some watered-down direct commission program and had probably never even been yelled at by a DI and as a result he knew of nothing about the USMC way of life – nothing about what I had been through. To put it quite simply, he couldn't relate.

I told him some of my struggles and even though I don't remember all of what he said to me, I do remember that he was very unsympathetic to my plight overall. The gist of his response was something like, "well, son? What do you expect? This is the Marine Corps. It's a rough world. Stay with God and pray."

Yea, sure, Sir. That will do me a lot of good.

There I was, having to watch my own ass, literally, and not even the damned chaplain gave a shit. Nothing more came from that meeting with the chaplain; however, unbeknownst to me at the time, he would end up getting me set up with a Navy shrink later – months later.

I went back to my platoon as if nothing had happened and continued with everything like normal. I heard nothing about these events for several days, but it was on my mind heavily. There was basically nothing I could do, but I knew I could not just let it go.

I was not in the place where I could just go and talk to someone who would listen or to request a change of unit or assignment as it just didn't work that way. We had the chain of command and that was law in the USMC. A Marine never was to go above the chain of command for anything and doing so without the proper permission could - and often would - result in punishment.

So, all I could do was go first to Corporal Jacobs, my squad leader, and try to request something that may help. Corporal Jacobs tried to be sympathetic and motivate me like Sergeant George had done in ITB, but things were a bit different than they had been in ITB. He told me I was a good Marine and that I was really good with the rifle and that I was valued in the platoon. I was not trying to hear any of it and eventually he sent me to my platoon Sergeant who gave zero fucks.

My platoon Sergeant was a new Staff Sergeant that I didn't know very well, and he ended up being a total asshole who basically ridiculed me the whole time I was speaking to him about the issue. He did not want to deal with me and considered me a problem meant for someone else. His solution to the whole thing was to remove the "problem" which he promptly did several days later by taking me from third platoon and putting me in second.

Once in second platoon, I again tried to go up the chain of command but was stopped by my new Sergeant who, while not a complete asshole, was generally unsympathetic and did not allow it to go any further. He basically told me to suck it up and deal with it. I was then just going to be stuck in a new platoon with no avenues for help.

That was bad for me and basically meant that everything which happened and all that I had brought up to both my Sergeant and Staff Sergeant in third platoon would be swept under the rug and treated as if it never happened. In second platoon, they knew essentially nothing about what had happened and moreover did not care. It was like I was a new boot all over again. I had a new platoon Sergeant, squad leader, and a new fireteam leader. My fireteam leader was Corporal Frye and he was a total dickhead. He was one of those "se-queer -ity force" guys and started in with the abuse as if I was a new Marine. I was at my absolute breaking point after everything that had happened and then also having to deal with the bullshit in second platoon. I felt like there was no way out, so I did the only thing that I could do – I continued with the job and tried to forget everything.

Eventually days and then weeks passed, and I began to acclimate to second platoon. At least I had the company of Castillo who, while in a different squad, was always a buddy to talk to even though I didn't speak of anything that happened in third platoon. The events of that day began to drift further out of my mind, and I again focused on surviving and doing the day-to-day tasks an infantry Marine does. Patrols, shooting, more patrols, more MREs, sleeping in the dirt, and of course … more patrols.

Corporal Frye made my life a living hell for the first month or so and I remember one night when we were on a patrol, he got so deep onto my case that I almost wanted to shoot him. I remember crossing a road during the patrol and being screamed at by him because there was a military vehicle driving about 100 yards away. He was screaming that I could have been killed, that

I was a piece of shit, that I was stupid and everything else under the sun. I remember thinking, *Motherfucker! That Goddamned vehicle is a whole football field away from me! I'm not a child; I know how to cross a fucking road!*

I remember him screaming at me, "I know why you're in this platoon. I know you were a piece of shit in third platoon and that's why they moved you over here!" Obviously, the higher ups were not giving the Corporals the real reason why I was in second platoon, so to everyone else it just looked like I was some problem Marine.

Each day I was losing sight of why I was there and what I was doing. I could no longer reconcile with myself and my decision to join the Marine Corps and everything they had taught me in bootcamp began to seem like complete bullshit. Honor, courage, commitment? Where the hell were all these values in the Marines to the left and right of me? Never leave a Marine behind… well, what happened to me? It was just total hell, but despite all that, I continued to grind my way through and eventually found myself on a naval ship which docked at a Naval base in Virgina.

On the ship I was tasked primarily with guarding the tons of USMC military gear – vehicles and everything in between – which we used, and which was staged in the hull of the ship. I was a broken person by that point – angry and bitter. I remember looking at myself in one of the head mirrors and thinking to myself that I looked like total hell. I had a half dead look on my face and realized that I had achieved the 1000-yard stare.

Down in the hull on one of my 8-hour watches, I screamed to

myself and yelled obscenities at no one. I remember being caught by two enlisted Navy girls who stared at me like I was crazy. Once they saw that I had noticed them, they eventually mustered up the courage and asked me, "Were you in Afghanistan or something?"

"I don't want to talk about anything. Leave me alone." They could see that there was something wrong with me; I knew it too and as they walked away talking to each other under their breath, I almost burst into tears.

Eventually, we bused back to Camp Lejeune and back to the barracks. I was put into a new room with a guy I really didn't know too well and didn't like that much. He would make a point to try and act tough and embarrass me whenever he was on the phone with his girlfriend, which was pretty annoying. Outside of that though, I began to put the whole situation which happened behind me and was even getting close with my platoon and fire-team leader. At that point I was a Lance Corporal, and the boot hazing that I experienced when I was first put into the fleet was all but over.

Things seemed to be getting back to normal, and I was even coming to terms with everything that happened; and even though I was bothered by it all and the other Marines treated me sort of differently because of it – bumscoop and what not – I tried my best to carry on. I felt like I had been put through the ringer with everything that had happened including the changes to a new platoon which was essentially like being a boot all over again, but damn it – I had done it!

I ended up buying a little used car and spending a lot of

time off base just driving around aimlessly and doing my best to stay away from the base as much as I could. I still hung out with Castillio and a few other Marines, but I was primarily spending a lot of time by myself at that point and withdrew into myself even further as time when on. About a month or so later, everything would change for me in the worst possible way, and once again, I had no control over it.

CHAPTER 9

THE CHASTISING

A NAVAL APPOINTEMENT

One random day as my squad was hanging out in one of the barracks rooms sort of joking around and having a good time, my squad leader entered and told me to come outside with him. I had no idea what was going on and had by that point tried my hardest to put everything that happened behind me. As I walked outside with him, he informed me that an appointment had been set up with a naval therapist and that I was to report to a specific area in the next thirty minutes. I followed his orders and went to the designated place I was to report to.

In my mind I felt anxiety about going to that appointment because I knew how it would be looked upon by my command. I had been unaware that an appointment was even made, and I am assuming it was a result of my visit with the chaplain which had been months before. I remember sitting in the waiting room area and filling out an intake sheet which had all the symptoms I "may" be experiencing next to a checkbox as well as a space for me to write my reason for being there.

I was very hesitant to be forthright on the sheet and as a result, only stated that I was there for "issues" I was having. I knew that if I said certain things they would greet me with punishment including solitary confinement and plethora of other medical practices that I was uninterested in partaking in not to mention ridicule by my command for being "weak". I did not want to be discharged on psychological grounds or have a black mark follow me around, but at the same time, I couldn't be too lenient and let what happened go, and I knew that it may be my only chance to try and get out of the mess I was in. I had completely put out most of what happened and had just planned to tell the guy that I was having difficulty "coping" in my unit.

Eventually I was called back into an office where I was greeted by an enlisted Navy corpsman. The guy was certainly not a doctor and probably had little training in terms of mental health or medical treatment. He asked me several questions about what I was feeling and what was going on with me and other customary type "probing" questions to try to ascertain my issues. I was very careful in how I worded things and kept it simple with, "I can't handle it or do it anymore." and "I'm having issues in my unit...". He seemed confused and kept pressing me for more, but I wouldn't budge. I told him about the panic attacks and anxiety I was having but did not relate it to any incident. Looking back, I most definitely should have!

After a half hour or so, he asked me one simple question, "are you wanting out of the Marine Corps or do you want help to stay in?" I already knew after everything that happened, I was wanting out – plain and simple. In my mind I had enough and saw enough; there was nothing more to be gained for me by staying in any longer. "Are you giving me an option?", I questioned, almost feeling as though he was joking with me. He paused and stated, "I'm asking you if you want to stay in or not?"

"Well, if you're giving me the option then I would like to get out.". He nodded and said, "Okay, I'll let the doctor know."

It was as simple as that.

I *never* saw a doctor which I found unusual given they were going to discharge me for some reason – most likely medical in nature. How was I not going to see a doctor? I probably should have pressed to speak to the doctor but for the sake of expediting the process, I didn't. What they recommended, I have no idea. They could have written I was a psychopath for all I knew. They never told me, and I still don't know to this day.

After several moments, the Navy Corpsman returned to me and said that the recommendation had been approved by the doctor, and that the recommendation was faxed over to my unit. That worried me because I could only imagine what they were going to do or say once they saw that recommendation. The Navy guy even told me, "Now listen; your command is going to put up a major fight over this. They are not going to take this discharge lightly."I listened as he continued, "they are going to confront you and do who knows *what* but listen to me: don't do anything to get yourself in trouble because eventually you'll be out, and you won't have to deal with them ever again."

He seemed very concerned for me when speaking about how my command was going to handle the discharge recommendations and all at once, my anxiety rose to a new and frightening level. They already didn't like me over in that company office and I knew that the recommendation was not going to sit well with them. I remember him telling me, "...watch your P's and Q's."

I knew I would have to bite my tongue, dig in my heels, and deal with whatever they were going to throw at me for however long it was going to take me to eventually be discharged. "I understand", I said back to the Corpsman.

After I left the appointment and returned to my barracks, and I mean immediately upon my arrival, the word was out and it was promulgated among the ranks, spreading almost as fast as a California wildfire. I was wondering how the fuck everyone knew about it already. It had only been 15 minutes since I left the appointment and somehow everyone and their mother knew what was happening with me.

The fact of the matter was that none of those Marines *actually* knew anything about what had happened which led to the Navy's recommendation for my discharge and anything they were being told was likely just vicious "bumscoop". That didn't matter though because to them I was just a piece of shit, a coward, and a pussy and I was going to have to get really used to being treated like one over the next few months.

Corporal Frye approached me and barked, "Verlice, fucking seriously?" I said nothing as he stood there looking at me like I was some type of mutant. "Take your rifle to the armory and turn it in." They were already making me turn in my rifle so immediately in my mind I thought, *am I getting out that fast?*

I had no idea how long the process would really take, but I was lulled into a false sense of security with the command to turn in my rifle. I falsely assumed that the whole ordeal would be over in a matter of days or maybe a week or two. Little did I know it would take many months; months filled with so much torment and abuse I would never have been able to imagine in my worst nightmares.

After I turned in my weapon and returned to the barracks, I realized that nobody would look at me or speak to me. It was almost as if I had the plague and I was to be avoided at all costs. Even Castillio wouldn't speak to me anymore – Castillio who had been a good friend of mine for over a year. Perhaps they were threatening him, saying things to him like, "if you talk to *Verlice* anymore, we will fuck you up just like *they've* done to him." I can't say I blamed him, but I can say if the shoe had been on the other foot, I would not have done the same. Castillio was a brother to me and would have been one to the end no matter what type of threats they would have laid upon me.

Over the next few days, I was surprised to see that First Sergeant Kellie barely spoke two words to me. I had figured he was going to let me have it for the Navy's recommendation; however, he did not...not then anyway. I really did not know what was going on with my discharge recommendation and was completely in the dark about it as nobody would speak to me. The truth of the matter is that nothing was going on with it. First Sergeant Kellie refused to look at any of the paperwork or get the ball rolling in any way for many months. He had different plans for me, and they did not include allowing me to be discharged free and clear. He was going to make sure that I left the Marine Corps with scars – both mental and physical.

ATTEMPTED DROWNING

One morning, my platoon was informed that we were going to take part in a water training exercise. The training was very specific and was something that they called "EGRESS" training. The training entailed learning to perform an emergency exit in a helicopter while submerged deep under water. The training was part of a workup for the next deployment in which the unit was going to be part of as the assault element of a MUE. A MUE was essentially a deployment where we floated around on a huge aircraft carrier and pretended to be at war for upwards of eleven months.

To me the whole idea of a MEU was almost like we were in the Navy and on one of the Navy's deployments, which was something that didn't interest me in the slightest. I had joined the Marine Corps for one reason only – to go to war! Regardless, the training was required for the MEU workup and everyone in my platoon had to go through it since we were the designated air element of the battalion and would need to know what to do in the event that the helicopter transporting us went down in the ocean.

Even though I knew I was going to be getting out of the Marines at some point and was not going to be participating in that bullshit deployment, I still had to do the training. I thought it was stupid and pointless, but I obeyed and followed the orders given to me and as such took part in the EGRESS training along with the rest of my platoon. I walked over to the designated area early that morning, feeling like something bad was going to happen. Once inside the building, I could see everything they were wanting us to do.

Inside the building was like a huge swimming pool, only much deeper. There were a couple different sections within the large pool which included different "stations" including a chair that flipped around underwater and a large helicopter body which was raised by a crane and then dropped into the water from where we were then supposed to escape. I performed several of the tasks including being submerged and flipped around in the underwater chair and some other simulations before taking part in the helicopter drop and escape.

The main event entailed being strapped into the helicopter body which was then dropped into a 40-foot body of water. It would spin and roll around in the water at which point all the Marines would have to remove their harness and break out the windows or doors and swim to the surface. The kicker to the whole training and the only thing that made it possible was the oxygen canister which we were to use during the escape. Every Marine was given a small canister of oxygen which was kept in the pocket of their trousers and contained only enough oxygen for three or four breaths. The oxygen canister was needed for the exercise because the depth of the pool would make it near impossible to escape the helicopter and break the surface without needing air.

They had us practice with the oxygen in shallow water in order to acclimate us to using the canister. I was having trouble using mine for some reason and could not get any air to come out and into my lungs. I'm not sure if it was malfunctioning or not but I brought it up to one of the Sergeants there who then briefed the CO on it.

My CO was Captain Hester and he had taken over for my original CO many months prior. Captain Hester was a tool in every way; the only thing he cared about was PT and the Warriors Ethos Handbook 10-12 (which detailed everything a Marine infantryman needed to know and was something nobody but him gave a shit about). Captain Hester was very close with First Sergeant Kellie, and I am quite sure that he was briefed on the situation regarding me and was most likely culpable in a lot of what happened moving forward.

The Sergeant came back and then instructed me to begin doing all these exercises without the oxygen tank. "If you can't use it, then you won't have it!" he barked, snatching the oxygen cansiter from my hands and throwing it in a bag behind him.

After having the oxygen taken from me, I began to feel as though something terrible was going happen, like they were going to force me to do the whole thing without oxygen. Let me be clear, it was *not* one of those training events (like one sees with Navy Seals) where they forcibly try to drown a person or keep them under for an extended period of time. That was not the point of the training, and I must underscore the fact that EVERY other Marine there was given oxygen. It clearly put me at a very unfair disadvantage and was reminiscent of the time the DIs forced me to go into the gas chamber for a second time without a functioning gas mask. It's all very ironic to me now – how everything mirrored bootcamp just on a much more dangerous level.

It was at that point that I knew they were trying to pull something on me, and I felt like it was going to be time for me to push back a little. Nevertheless, I was in fact forced under the water and held for long periods of time without oxygen as everyone else went about their business and ignored what was going on with me. It was to the point where I was beginning to black out and was on the cusp of breathing in water. After several iterations of being forced under, I stated that I could not do anymore and that I was finished.

The CO, who had been standing nearby and most likely watching it occur, made those involved in the activity stop at onece to which they promptly did. Anytime an officer was around nothing "bad" was really supposed to occur, even though they knew it happened and, in some cases, even authorized it, they still couldn't "see" it lest they be held liable for any damages or injuries. I was made to report to Captain Hester who told me, "Okay, Verlice. Come back tomorrow and you will do it again."

"Aye, Sir.", I responded shakily.

I rushed out of the building shaken and felt defeated by what had just taken place. I truly felt as though they were attempting to cause bodily harm or even worse – as retaliation – for the Master Sergeant Sneed and First Sergeant Kellie situation. By that point, First Sergeant Kellie was beginning to rally everyone in the company against me and as a result Sergeants and low rankers alike were all beginning to "get some", and it was almost as if an "order" had been given to fuck me up and harass me given every opportunity they could. I was okay with going back the next day even though I didn't want to after what had occurred. Outside, a Sergeant intercepted me in front of the building and ordered me to get into his truck. He said he was going to drive me back to the company office but didn't specify why. I knew in my gut something bad was going to happen when I got back to the office. It was the first time I met Staff Sergeant Billings.

STAFF SERGEANT BILLINGS

Up until that day, I had no idea who Staff Sergeant Billings was and had expected to have another "meeting" with First Sergeant Kellie once I got back to the office. Staff Sergeant Billings was not in my direct chain of command nor was he even in a rifle company; he was a weapons company platoon Sergeant – some machine gunner or mortar man in charge of weapons platoon. As stated before, the rifle platoons and weapons company did not hardly mix in training, the field, or anywhere else, so it was not usual for a weapons company Staff Sergeant to have direct daily dealings over a rifleman.

Staff Sergeant Billings, it turned out, was a close buddy of First Sergeant Kellie and in fact, almost seemed to be his little side kick. Moving forward it always seemed as though the two were connected at the hips and wherever First Sergeant Kellie was, SSGT Billings was right there next to him ready to echo all his threats and taunts. Looking back, I have the feeling that First Sergeant Kellie had told SSGT Billings, "Hey, I have this guy here that we've been fucking up. If you want in, come and get some!"

and "get some" he surely did. Anytime the First Sergeant laid on a beating, SSGT Billings was there to give me double.

Staff Sergeant Billings was a cruel individual with a permanent frown glued upon his craggy face. He looked like a goon with crazy, sunken in eyes and gaunt, hallowed out cheeks. He was lanky and lean and had very jagged, almost sharp pointed teeth. He looked like a reincarnated Nazi solider from WWII and had the nasty attitude to match it. The guy was plain and simply a total asshole.

Once the Sergeant got me back to the company office, he told me to wait outside for further orders. I stood outside the back door for several moments not knowing was to come and after a few minutes I heard my name, "VERLIIIICCCCEE..." being barked from inside.

I knew in an instant, based on the tone, that it wasn't going to end well for me. I hesitated for a moment but ended up walking in and once inside, I realized that it had not been First Sergeant Kellie who was yelling for me, it was that Marine who I did not recognize. It was Staff Sergeant Billings.

With no words or introduction, he pounced on me and immediately started assaulting me by way of strangulation. He was screaming at me to go back to the water area to continue doing the exercise I had been doing before being driven back to the company office. Even though Captain Hester had told me to come back the next morning, which I had agreed to, Staff Sergeant Billings took it upon himself to take matters into his own hands. At that point, I could honestly say that if he had got me back into the water, I may have been drowned that day and there was no way I was going to let that happen. I started to tense up and resist his strangulation but did not fight back. I did not fight back because as an E3 it would have been an instant court martial for hitting an E6 and nothing I said would have exonerated me. There was nothing I could do but take the beating. As I was strangled, I found myself going in and out of consciousness several times and each time there seemed to be a moment in time which was lost.

When I came to for the last time, SSGT Billings was dragging me out of the company office doorway and attempting drag me outside. I was putting up as much resistance as I could because there was no way that I was going to let that guy get me back in the water. He was screaming, beating, and calling me all sorts of obscenities as he tried to get me out the door.

Eventually he got me outside where there were plenty of witnesses to see what was going on but that did not slow him down. He still had me by the throat and was essentially dragging me inch by inch because I was putting up quite a bit of resistance at that point. Why couldn't an officer or Sergeant Major happen upon us in that moment and put a stop to the assault? Regardless, I managed to get, "STOP!", out before I think he realized that someone may eventually see what was happening and as a result, let me go. I fell to my knees coughing and gagging as I attempted to get air back into my lungs. SSGT Billings had disappeaered, I'm assuming, back into the company office somewhere.

When I stood up, I was completely out of sorts and in a state of total confusion – fog of war, if you will. I was discombobulated, disoriented, and disheveled. My cover had somehow fallen off my head during the assault and I had no idea where it was. I was standing outside without a cover on my "grape" and that was something that would quickly be noticed by passing Marines as one was never to be outside in uniform without a hat on their head. I was still relatively calm at that point, but I could feel myself slowly coming apart at the seams.

COROPRAL HENDERSON

As I stood there, dazed and confused, I heard a Marine passing by start to scream at me, "Bitch, where is your cover! Put a cover on your damn head!"

Great, I remember thinking.

now I'm going to get fucked up over my damn cover.

As I haphazardly tried to find my cover, I noticed a Marine walking down the sidewalk in front of the company office. He was a Marine that I did not know personally but had heard through the grapevine from other Marines who had known him. It was Corporal Henderson. Corporal Henderson was a notorious "boot hater" and the stories surrounding him involved hazing, assault, abuse, and everything in between. He was also in Weapons Company, like SSGT Billings, so I had not worked with him directly, but his reputation certainly preceded him as he stomped his way over towards me with a look of anger and disgust gripping his face.

I had just been choked out multiple times and now I was about to get another asshole ripped for not having a cover on my head while outside. As he approached me, I quickly spun on my heals and tried to figure out where to go. I didn't really want to go back into the company office because of what had just occurred but there was nowhere else for me to exit. I turned around and saw Corporal Henderson approaching. "What is going on with you?" he questioned. I readied myself for another ass beating.

Corporal Henderson approached swiftly, but he did not yell, and he did not scream; he did not put his hands on me in a violent way. "Are you okay…" he had to look at the name tape on my uniform, "Verlice?"

When I heard the compassion in his voice, I let go emotionally. Given his violent reputation, I had figured that he was going to beat the hell out of me; however, he did the exact opposite. When I started to let tears flow, he showed nothing but concern and grabbed me in an embrace and pulled me into him.

"It's okay, Verlice."

I couldn't believe he was showing me that type of compassion; of all Marines…Corporal Henderson! He whispered in my ear, "Me too, Verlice. Me too."

He had sympathized with what was going on with me and was in fact the only Marine in the unit who ever had. Other Marines passing by witnessing the event were laughing and snickering, completely ignorant to anything I had just endured. It was only Corporal Henderson who did not and instead showed some humanity.

He tried to question me to find out what was going on, but I was so incoherent and inconsolable that I was unable to get any audible words out. He just held me in an embrace as one Marine would do to another who was hurting. After several moments, First Sergeant Kellie emerged from the company office building and screamed for me to follow him. I had no choice but to listen, so I left Corporal Henderson, who still showed a major look of concern on his face as I walked away, and entered the company office once again.

SERGEANT MAJOR ARNOLD

I'm assuming the First Sergeant either saw the assault take place or someone else within the company office saw it and realized that something had to be done to address the situation before some higher up or officer witnessed it and then the problem became bigger. The First Sergeant escorted me upstairs and right into the Sergeant Major's office.

I should preface this by stating that it was incredibly rare and uncommon for anyone to go in front of the Sergeant Major and typically only occurred under serious circumstances -infractions and punishments, so to say I was nervous about going in front of him would be an understatement. The Sergeant Major was the highest-ranking enlisted Marine in the unit and the only Marine that held more power in the office would have been the Battalion Commander.

I was nervous and shaking as I was led upstairs and into the Sergeant Major's office, a place I *never* thought I'd be. Once inside his office, I stood at parade rest looking disheveled as I tried my hardest not to break down emotionally.

The Sergeant Major was like a character straight out of some movie or thriller novel. He stood about six foot eight and resembled John Wayne. He looked incredibly intimidating with his jutting underbite, His voice, like a powerful thundering boom, was even more intimidating and could easily scare the literal shit out of any Marine unfortunate enough to find themselves standing in front of him. I always use to think that it was for those reasons why the Marine Corps made him a Sergeant Major as only someone with looks that ferocious would qualify for the job. The guy definitely looked the part of US Marine.

As I stood there in front of him silently trembling, he looked me up and down and then proceeded by asking me what the problem was. I am not sure if the Sergeant Major had heard or seen - or both – what had happened to me downstairs, but I assumed he most likely did. Perhaps that was why I was in front of him because essentially, I had done nothing wrong, so to be punished for what had happened, I think, would have been a bit gratuitous. I tried to explain, to the best of my ability, what the problem was but was unable to speak frankly primarily because of the large disparity between ranks. I was too intimidated by both the Sergeant Major as well as the First Sergeant, who was standing right next to me, and as a result, I did not come forward with what had actually happened regarding SSGT Billings and myself. It probably would not have mattered anyway as I would most likely have not been believed.

Every time I tried to say something, the Sergeant Major would cut me off and order me to do something like stand more erect or speak louder. Eventually, I was able to describe somewhat of what had taken place in the water and why I didn't want to go back; I chose not to mention anything that SSGT Billings had done to me for a variety of reasons but primarily due to intimidation. While The Sergeant Major was stern with me, he seemed somewhat sympathetic to my plight and even offered some condolences.

"I understand. That is a legitimate reason and don't worry, you won't be made to go back after that!", he assured me.

In the Marines, the higher the rank, the more serious they *should* be about the rules of the Marine Corps - rules including no assaulting, hazing, or … yea, attempted drownings! At that rank, they are completely expected to handle situations by the book and that is exactly what he did. Looking back, I wish I would have spilled my guts about everything: Master Sergeant Sneed, the assaults, the attempted drowning, and everything in between, but until *you* stand in front of a high ranker like that *and* with the amount of brainwashing which I had been through, *you* just don't know what *you'd* do.

I remember he started going through "files" on his computer, mentioning things about what he saw written in them. "No issues in bootcamp, ITB, or your platoon, so I'm not really sure what your issue is here, son." he puzzled.

I tried to explain to him that I was in the process of being discharged because I got the notion that he was trying to also prevent that from occurring based on what he was telling me. I even tried, in desperation, to tell him that I lied during the MEPs process (which wasn't true), hoping that would give them a reason to remove me from the platoon or expedite the discharge process. None of that mattered to him one bit and he just looked at me and shook his head, "well, too late for that now, isn't it?"

The fact of the matter was that I was *not* some sub-par Marine who couldn't hack the infantry. He pulled file after file on me trying

247

to find anything to make me fit a certain type of Marine – the type of Marine considered weak or prone to mental breakdowns, but he simply couldn't find it because it wasn't there. I had a first class PFT score (290 out of 300), was an excellent shot, had never been dropped in bootcamp or ITB, and had no reprimands throughout my time in the fleet.

"You are the perfect Marine on paper, Son. Why are you in here standing in front of me?" Perhaps I should have said the real reason right there, but again, the higher the rank, the tighter my asshole got, so I stayed quiet.

I was holding back tears as he continued going through my files and mentioning my parents by name. He even called my mother by her first name and said, "what would Janet think about this?" At that point, all I wanted to do was go home and with everything I had been through in the past several months, hearing my mother's name just made me want to crawl into a hole and die. He looked at me as I stood there fighting tears and barked, "you know why you can't cry?"

I didn't say anything as he continued,

"it's because you are in front of another man and it's a sign of weakness. That is why you can't cry right now, Verlice."

I refused to let tears fall in that office, so I just stood there at the position of attention while he went on talking with the First Sergeant about what to do with me.

The meeting ended with the Sergeant Major giving the order to First Sergeant Kellie that nobody was to put their hands on me again or assault me in any way shape or form or there be serious consequences. First Sergeant Kellie even agreed and assured me again after we left the Sergeant Major's office that nobody was to assault me and if they did, I was to report it to him immediately. That was sort of ironic given *he* was the one who instigated most if not all the physical assaulting.

After that day, I was finally removed from the rifle platoon and put, temporarily, in headquarters company (HQ), which was still part of the unit with the same leadership ran out of the same company office. So really, it didn't do me any good. I did, however, feel a bit better afterwards because I thought that no more abuse would occur and that I would just be discharged after a short while, allowing me to finally forget about everything that had happened. Unfortunately, it did not happen that way.

The next week, the First Sergeant continued to harass me, mainly on a mental level, all while having me doing ridiculous tasks in "HQ" company. I was also taking flak from every other enlisted Marine in my company for the move to "HQ" and was basically labeled a "nut case" amongst the ranks. Things were to the point where I had to resort to sleeping in my car at night just to avoid being harassed or targeted by everyone I came into contact with. I did that for approximately three months or so until I was given a different room in a different barracks. The First Sergeant even knew about me sleeping in my car and did not address it; it was exactly where he wanted me. It was essentially open season on Verlice.

In HQ company, the First Sergeant had me answering a phone – the company phone – all day long. The phone was hooked into the wall and never rang…not even once. The only time it rang was if *he* called it to say something like, "Verlice…fuck you," or any number of other insults, threats, or taunts.

After about a week or so in HQ company, I think he realized that I wasn't going to be doing or contributing much, and I was therefore moved into a platoon which was called "RBE" (remain behind element). That platoon consisted of any Marine who was being discharged for one reason or another - be it medical, mental, end of service. Essentially every Marine in the RBE platoon was left alone due to those medical purposes and because they were on their way out of the Marines…except for me! I was the only one from that platoon who was continuously singled out by Sergeants as well as the First Sergeant.

During that time, I made a good friend who was also in RBE. He was in a similar type of situation as me - although he wasn't being singled out the way I was - and had the same feelings about the Marines as me. I was not 100 percent sure why he was being discharged, but the story was that he turned a fully loaded rifle around on several Marines during a "melt down" he had while in the field. Whatever the reason, Garnett and I became quite close and spent a lot of our time off base doing whatever we could to keep our minds off the shit we were dealing with. Even though I was very close with Garnett, I never told him anything about what happened and why I was in RBE - not the real reason anyway.

Around that time, I was beginning to have more internal problems including nightmares, trouble sleeping, and anxiety. I was having panic attacks more frequently and struggling with suicidal thoughts on a regular basis. I remember one night I contemplated hanging myself with my belt on the pull up bars outside of the company office but decided against it and continued to wait for word on my discharge. I had heard nothing of my discharge situation since the Naval Appointment I had had a couple months prior and was beginning to think that they had other plans for me.

It was also around that time that, as a punishment, I was put-on 24-hour guard duty in the barracks every other day. I didn't mind it so much though because on duty I was left alone in a small office all day and it gave me time to go over everything in my mind. I did not try to think about the events too much however, since I found it hard to focus on it all while things were still going on. I knew that I had to get out of there before I started to think about everything in greater detail.

One day for duty, which always occurred with one higher rank, I was paired with Corporal Henderson. We talked together all day about what had happened, and I told him things about my plight – albeit not everything. He also told me things about his time in the Marines and that he had struggled with similar scenarios but somehow overcame it. He was a very nice guy, despite his reputation, and I held him in high regard and still do for that matter.

There were many other people like me in RBE – Marines who were getting out for similar reasons related to abuse and hazing. I was not the only one in that type of situation; however, I do believe that I was the only one getting violent treatment as I saw nobody else getting assaulted or heard anyone else talking about similar situations. As a matter of fact, the First Sergeant was viewed by almost everyone in the unit as a laid-back guy and most of the Marines referred to him as being "cool as fuck". Hell, I even thought that about him sometimes, because at times he could be, even with me.

But for reasons still unknown to me, he took a disliking to me and saw to it that I was assaulted on a regular. He would say things like, "You're trying to play the system", or "I'm gonna make sure you leave here broken."

I believe the only reason that I was even given the chance to be discharged in the manner to which I was going to be was because of President Obama's initiative to draw back the US Military during the time I was serving. That gave them a reason to give anyone – for any reason – an honorable discharge. I believe if that had not been the case, I most likely would have had to get myself kicked out on some bad conduct.

As the next month rolled on into November, and as the USMC birthday approached, I was given the duty of "usher" at the annual Marine Corps Ball which was held that year on a Saturday. I remember I had to get my Alpha uniform starched and pinned with my new rank since I had not had to wear it for well over a year. It remained at the dry cleaners for several days where I was to pick it up the Friday before the ball. Earlier that week before the ball, I was informed that I had an appointment to attend that Friday morning right before the ball would take place.

Early on that morning, I was to report for something called SEPs/TAPs and that was something which was required for any Marine separating; it was a class about what to expect once you are discharged such as assimilating back into civilian life. During that week, I spoke to the MO (medical officer) of my unit and informed him of the meeting and was told that it was fine to head over there early that morning without going to formation. That Friday also happened to be the day that there was going to be a large battalion exercise and since I was going to that appointment, I would not be taking part in it. Everything was approved; however, the First Sergeant had different plans.

CHAPTER 10

THE CRUCIFIXTION

SERGEANT BARRET

The evening before the battalion event, on the 8th, the platoon Sergeant of RBE, Sergeant Barret, informed us all that nobody in RBE would have to partake in the exercise due to medical reasons or appointments. It wasn't until he noticed me standing in the back that he called me out and informed me that I, in fact, would be the only one in RBE partaking in the exercise. His announcement threw me off guard given the fact that I did have an appointment the next morning, and because it was known and approved by the MO of the unit. I couldn't help but wonder to myself why the *fuck* he was singling me out, but I figured it was just because of the type of person he was. Sergeant Barret was a snake in every meaning of the word. He disliked me, which almost everyone did by that point, but he made it personal by calling me out the way he did. Looking back, I'm quite confident that it was part of a set up orchestrated by higher command, likely the First Sergeant and maybe even the CO.

I politely informed Sergeant Barret that I had an appointment scheduled for the next morning and would not be able to perform the activity. He seemed to take my explanation as some type of refusal to obey him and as a result, he got pissy and then proceeded to make me accompany him to the company office after formation let out. I was, again, wracked with nerves as the company office was a place of trauma for me by that point and I really did not feel like taking another beating.

Once in the company office, I was "greeted" by the First Sergeant who was less than happy to see me. He didn't put his hands on me then, likely because the CO was present along with him and would have been to risky to do so with officers present.

First Sergeant Kellie was in one of the worst moods I had seen him in thus far and he immediately started to yell and scream at me, acting as though I had defied some order given to me. *He* knew of my appointment and basically said, "fuck that, you're doing this tomorrow, I don't care about that appointment!" He then made me go in front of the CO, Captain Hester, who basically scolded me and said, "Verlice…cut out the bullshit."

'Excuse me!?', I screamed in my head.

What bullshit? You mean the attempted sexual assault by your Master Sergeant? Or maybe the strangulation and beatings by your Staff Sergeant and First Sergeant? Is that the bullshit you are referring to?

Of course I couldn't say anything like that to him, so I just stood there at parade rest and said, "Aye, Sir."

Once the CO was finished with his "speech", the First Sergeant screamed at me to get out of the CO's office and wait for him outside the door. Once outside, First Sergeant Kellie proceeded to mock me and tell me that I was going to be "fucked up" the following morning during the battalion exercise. I didn't know what that meant, but I didn't like the sound of it.

"Do you understand? You're going to be doing this tomorrow and trust me, it will be a memorable day for you."

I knew that I could not fight a First Sergeant's command or disobey any of his orders directly, so I just replied, "Aye, First Sergeant."

That was that; the plan was set.

I was going to do what he told me to do because I did not see any way to get around it. I was not going to be put in a position where he was going to be able to assault me again or make me look as though I was disobeying orders. I didn't know what he was planning for me the following morning, but whatever it was I knew it wasn't going to be good. I honestly think, looking back, that I was not going to come back from that exercise alive.

After giving me the command for the following day, he told me to leave the office, and I promptly did without hesitation.

Now where things get weird...

Once I left, Sergeant Barret followed me out and walked me back to the barracks. That was something that a Sergeant, especially

Sergeant Barret, typically NEVER did. It was the beginning, I believe, to their plan which they were setting in place. To what extent it was schemed, I'll never know, but I can reasonably infer based upon what happened next that there was in fact some type of plan which was going to exact some final revenge upon me.

Sergeant Barret walked me over to the smoke pit and gave me a cigarette. It was pouring rain at that point, and I was wondering why he had decided to follow me over to the smoke pit to "share" a cigarette with me. I wondered why all the sudden he was acting nice to me when he had never done so before. He talked to me like a friend - like a brother - and I ended up telling him a bunch of stuff that bothered me from my childhood and other events not relevant to the USMC.

I don't know why I chose to tell him the things that I did— I guess I was in a vulnerable state. I told him nothing about what had happened to me in the unit over the past several months and considering what happened the following day, I'm very glad that I didn't. He appeared to take pity on me and even gave me a hug, which I was not looking for in the slightest. He also told me he was going to hook me up with his "therapist" off base because he thought they could help me sort some things out. I did feel as though he cared… big mistake!

Finally, before he left me alone, he told me that I did in fact have that appointment the following morning and that I was to come down in my uniform instead of PT gear because he determined that I would no longer be taking part in the battalion exercise. He told me to, "…come down in your uniform so that you can go to that appointment, and I will go right now and tell the

First Sergeant, so he doesn't put your ass on blast in the morning."

Ass on blast…

I can still hear him saying that to this day. He left and walked towards the company office and I assumed, ignorantly, that he was going to do what he said. I returned to my barracks room feeling somewhat relieved.

I ended up waiting until 2 or 3 in the morning for him to text me or call me telling me that the word from First Sergeant was "yay" or "nay". I had a bad gut feeling that it wasn't going to fly with First Sergeant Kellie, but I thought, *no news is good news*. I couldn't have been more wrong. I eventually fell asleep and woke up around 5:30 AM; the world was still dark and cold, and it was about to turn even colder.

BETRAYED

I remember being petrified to go to the morning formation because I just knew deep in my gut that something bad was about to happen. When I got into formation, I noticed something odd: there were no officers around. Typically, there would be an officer somewhere during these morning formations - either a 2nd lieutenant or maybe even the CO – and since it was the day of a huge battalion exercise which included all the officers, it was very strange that none of them were present. Looking back, I think that was done on purpose; I believe they knew what they were going to do and wanted to minimize the risk of being caught.

I was sort of hiding in the back of formation trying to stay out of view of the First Sergeant, when I suddenly heard the familiar screaming of my name, "VERRRRLLIIIICEEEE!"

It was First Sergeant Kellie! Somehow, he already knew about everything even though he had not laid eyes on me yet.

sort of weird, right?

I had no choice but to head up to the front of the company to meet my maker. As I was walking up there, a couple of the Sergeants began playing fuck-fuck games with me such as making me get back, come back, get back, and so on, which was quite humiliating in and of itself.

Eventually, I made my way up in front of First Sergeant Kellie who was standing there looking as though he was about to eviscerate me with his eyes. There I was standing in front of the entire Alpha Company about to be made an example of and there was nothing I could do about it. The only Marines present in the front of the formation were First Sergeant Kellie, Staff Sergeant Billings, and Sergeant Barret. I found it odd that Staff Sergeant Billings, the Marine who had attacked me by strangulation a couple months prior, was standing there when he had no real reason to be. It was almost as if he *knew* what was going to go down that morning – and I'm sure he did.

"I fucking told you to come down in PT gear, Verlice!", barked the First Sergeant angrily. I said nothing and could only look at Sergeant Barret who stood next to the First Sergeant. I said nothing, but I begged him with my eyes; I begged him to say something to help me out. It was like I was saying,

Mother fucker! Tell him what you told me last night. Save me, man!

Sergeant Barret only looked at me with the coldest glare I had ever seen – a true psychopath – and simply shrugged his shoulders as he mouthed the words -

Oh well!

I knew at that moment that I had been duped – set up – and no matter what I did from that point forward, I was going to be screwed. I had decided, since I was already standing there, that I might as well take whatever was going to come. The only thing I *knew* was that I could *not* fight back, or they could have me arrested and court martialed. The world around me got quiet as I buckled myself in for one of the worst beatings I had ever received.

As the ass-chewing commenced, Sergeant Barret eventually disappeared as did the rest of the company. I'm sure everyone in the company had a good view of everything happening, but somehow, I was blind to it all. At that point, I had completely shut down and went into autopilot. The only thing that I said from there was either, "no, First Sergeant" or "yes, First Sergeant."

There was no point in trying to explain anything to him; it had been preordained –planned out – and they knew exactly what I was going to do and *exactly* what they were going to do, and nothing could have put a stop to what was about to occur.

I saw Staff Sergeant Billings glaring at me from the corner of my eye as the First Sergeant continued to scream at me. The guy was practically foaming at the mouth, just ready to get his hands on me when he got the chance. He had been itching for it since the last time we had met.

"Good, let's drag him into the laundry room so nobody can see us kicking his ass", commanded the First Sergeant.

The laundry room, where they intended to take me, was inside the ground floor portion of the barracks and was isolated, free of any cameras or people, and relatively private. The perfect place for a beating of the century.

Staff Sergeant Billings swiftly grabbed me by the throat, as he had done before, and the two of them dragged me into the laundry room of the barracks. I put up no resistance. Once inside, SSGT Billings threw me as hard as he could into one of the washing machines and I quickly caught my balance and stood straight up.

The two then took turns holding me and beating my ass for some twenty minutes. One would restrain me in a choke hold while the other beat and then they would switch off over and over again. The ultimate tag team! I could not do anything to defend myself; I knew that's what they wanted. They wanted me to swing back so that they could have a verifiable reason for beating my ass. I wasn't going to give them the satisfaction. If I was going to die, then I would die having not said a word and having not thrown a single punch.

They eventually took turns body slamming me as hard as they could onto the ground. It was not wrestling either; it was real… and it was painful! I was again choked in and out of unconsciousness by Staff Sergeant Billings as I was simultaneously beaten by the First Sergeant. I cannot remember all of what happened as there were spaces of time, or gaps, that I simply cannot remember due to being in and out of consciousness. I do remember them forcing certain liquids into my mouth and attempting to shove random articles of clothing down my throat. The whole time I didn't shed one tear or even wince in pain. Looking back, I'm not sure how I was able to do that. I guess I had just completely shut down.

I vaguely remember requesting Mast as the assault was taking place and it was promptly denied by First Sergeant Kellie who simply screamed, "FUCK YOU!" Requesting Mast was something afforded to every Marine under the UCMJ and was put in place for situations such as what was happening to me at that moment. Requesting mast allowed a Marine to go over the chain of command in the event of a serious situation being swept under the rug or being handled incorrectly. Typically, one would request Mast to go over their Sergeant or Staff Sergeant and speak to the company First Sergeant or CO.

My case, however, was a bit unique in that it was my company First Sergeant who oversaw the assaults. I also feel quite certain that the CO was complicit too and was at least aware of what was taking place to some extent. I would have had to request Mast to speak to either the Battalion Sergeant Major or the Battalion Commander, which would have been unusual for anyone requesting mast. Obviously, he was not going to let that happen as it would have been game over for both First Sergeant Kellie and SSGT Billings had either the Sergeant Major or Batallion Commander knew. Unfortunately for me, the request was promptly denied. That denial in and of itself was illegal on their part – according to the UCMJ.

I was again choked in and out of unconsciousness by Staff

Sergeant Billings and during that last episode of strangulation, we somehow ended up back outside in front of the barracks. I have no idea how we did, but we did. It was still dark at that point, and it appeared as if no other Marines were in the area – convenient! I was beginning to lose my cool a little bit and even attempted to get away from Staff Sergeant Billings who had briefly loosened the grip he had around my neck. I took one step away from him and he quickly grabbed me, picked me up, and slammed me onto the concrete as hard as he possibly could.

I slammed onto the ground with incredible force and felt every bone in my body shift and crack. It was almost as if I had a chiropractic adjustment right on the pavement. My right shoulder made a loud, greasy popping noise and was twisted almost 360 degrees out of the socket. The pain seared through my body like a hot knife and was so intense that I almost vomited. Somehow, once I shifted my weight a bit, my shoulder popped back in place. I could tell that it was damaged though, and it hurt for the next few months; however, I never received medical treatment for it.

Just another gift from the big green weenie.

I lay on the ground for several moments trying to compose myself, letting the pain from my shoulder subside before figuring out what my next move would be. "Yea don't act like I hurt you, bitch!", screamed Staff Sergeant Billings as he towered over me like Godzilla.

Somehow, I managed to get to my feet and back to the position of attention as the two just stood there glaring daggers at me. I was put back into another chokehold by Staff Sergeant Billings and that is when First Sergeant Kellie pulled the knife out on me. He produced the knife from his pocket; it had a six-inch blade on it and was razor sharp. He thrusted it towards me in an attempt to scare me and blood was drawn. As I twisted from Staff Sergeant Billing's choke hold, I could feel the knife slicing through my right arm. Blood flowed from the inside of my arm down my chest where it pooled in my trousers.

First Sergeant Kellie then put the knife to my throat and pressed it hard but didn't cut. I don't remember if he said anything as I was too overtaken by pain and lack of air to remember, but I remember the look in his eyes. It was homicidal rage, and I knew the guy wanted to kill me right there, and had I been in a more private area where nobody could have seen, I'm quite sure he would have.

After several moments, he turned the knife around on himself and urged me to grab it from his hand.

"STAB ME, VERLICE! FUCKING STAB ME!"

It was clear what he was trying to do; he was trying to have me reach for the knife so that he could either kill me in "self defense" and provide a reason for the beating they gave me, or court martial me for attempted murder. Had I done that, those two would have backed each other up all day long and anything which may have went to court would have been ruled in their favor. There are people who, having been pushed as far as I was that morning, may have reached for that knife just out of pure desperation. I just stood there gripping the Staff Sergeant's arms to try and get breath and did nothing at all.

As I stood there for several moments, feeling warm blood drip into my trousers and down my leg, I wondered how it all would end. Still, I said nothing and in an instant, First Sergeant Kellie and Staff Sergeant Billings disengaged and disappeared, leaving me by myself where I stood. Everything ended as quickly as it had started, and I found that I was in complete shock and disarray over the event. I hurried back up to my barracks room where I checked the inside of my upper right arm, which had a three-inch gash and blood trickling out. I cleaned it up and did my best to stop the bleeding. Obviously, I was not about to get any attention for my shoulder or the cut on my arm, but I certainly knew something more would come from what had just occurred.

THE BRIG

I'm not sure how much time had elapsed from when I was assaulted to when Sergeant Barret came into my room to get me, but it couldn't have been more than twenty minutes. I was too out of it and quite frankly in shock to even grasp the concept of time at that point. My shoulder was in pain and felt as if it didn't have full range of motion, and the cut on my arm, while not bleeding, was still throbbing. Honestly, the injuries were the last thoughts on my mind becasue I heard Sergeant Barret barge into the room, breaking my solice once again. He told me to follow him to his car because he had been instructed to drive me over to the base naval hospital.

The news was good, in a way, because I had a plan to tell everything that had happened to me to someone over in that naval hospital. The Marines didn't give a shit, but the Navy surely would have. The Navy was a different entity and were more apt to listen to grievances, especially one of this magnitude, so I knew that going to the hospital was my best bet at a fair fight. I surmised that if I had told them everything that I was going to, an investigation most likely would have been underway, and perhaps even a court martial, on both First Sergeant Kellie and Staff Sergeant Billings.

On the drive over to the hospital, Sergeant Barret sighed and said something to me along the lines of, "I can't believe this, Verlice", or "do you know what is going to happen now, Verlice?" I felt so much disdain for him after the backstabbing he had done to me, that I didn't even bother to say anything back to him or even address him by his rank. I simply stayed silent, and he didn't press me any further. I think he knew how bad the situation was and did not want to make the situation worse by giving me an ass chewing for not answering him by his rank.

He had driven me all the way to the hospital, and we were about to park and head in when his phone rang. It was clearly someone from the company office telling him to promptly take me back.

You've got to be kidding me

He quickly started to drive out of the hospital parking lot and back to the company office where my attackers were waiting. I had no idea what they were going to do to me once they got me back in that office and to say I was worried about it would be an understatement.

Once back at the company office, I was escorted to the Medical Officer's office where I was greeted with concern. The MO of the unit was basically a shitbag; he did not care about his duties or any of the Marines...at least that is how it appeared from all the interactions I had had with him. "What the hell happened, Verlice?", he said with a sincere level of concern in his voice. By that point, I was beginning to get angry about what happened and even yelled, "KEEP THOSE TWO PSYCHOS AWAY FROM ME!"

He seemed confused; he clearly had no knowledge of what had taken place that morning.

"What do you mean?", he questioned.

"Those two physically assaulted the hell out of me, Sir!"

I was trying to tell him what they did, but he kept cutting me off and failed to press me for further details. One would think the MO of the unit would want to know more, but he clearly did not.

The MO ending up leaving the office, presumably to talk to the First Sergeant who was in a different section of the company building, and then came back into the office again after several moments to relay information to me. I remember, vividly, him saying, "What if I told you to just go back to your barracks room and forget all of this happened?"

It was clear to me that they were trying to sweep what happened that morning under the rug. It was also clear to me that my command did not know what to do or how to handle the situation and were most likely shitting their pants at the possibility that I may end up telling someone – someone who might actually listen – what had happened. "Sir!" I exclaimed, "I really think I need to go to the hospital."

I remember the look on his face turning from one of concern to that of disbelief. It was like he knew something I didn't; it was almost as if he was saying, *dude, you just fucked yourself.*

"You, uh…want to go to the hospital?" he stuttered.

"Yes, Sir."

He let out a little irritating huff of disappointment and then left the office again. He must have went back to where the others were and told my First Sergeant or the CO that I wanted to go to the hospital. They clearly told him no because when he came back to where I was waiting, he told me, "Well, they're actually going to put you in the brig."

I remember that hitting me like something completely out of left field. I remember thinking, *The Brig! For what?*

I didn't do anything which warranted going to the brig that morning and was shocked to even hear him tell me that they were going to lock me up.

"Are you serious?", I asked in a tone which conveyed the utmost defeat.

"Unfortunately, yes."

He looked at me as if the whole situation was "out of his control". The MO was a full-fledged doctor who had been to medical school and everything. He seemed young to me, probably no older than 30 years old max, so he most likely had never been a doctor in the civilian world. He was a naval doctor attached to my Marine infantry unit and was likely close to the end of his tour of duty, so none of it probably mattered much to him anyway. I didn't know his story or what his deal was, and while he seemed concerned for me as a human being, he certainly did not do anything to help me in the situation I was in. He had the power to trump my command and tell them, "I think this Marine needs to go to the hospital to be checked", but he didn't. He did nothing to save me, and in my opinion, he failed at his job.

Even if I *had* done something which warranted going to the brig, the MO should have still requested that I go to the hospital first to get checked out simply on the basis that a major assault had taken place. The bottom line was that *they* knew what would have happened had I went to that hospital, and *they* did everything in their power to make sure I spoke to nobody (who would care) about the incident.

So, they put me in the brig that Friday and there was nothing I could do to stop it from happening. Before sending me, I was briefly looked over by a Corpsman who did nothing but ask me if I was homicidal or suicidal to which I answered no. There was no way I was going to be put in the brig with a homicidal or suicidal label hanging over me. It would NEVER have been met with sympathy.

Before being driven to the brig, *and let me reiterate that I was never formerly charged with anything, arrested, read rights, or anything that they should have "legally" done*, they sashayed me around Alpha Company clad only in a pair of underwear in one final act of ignominious display. As humiliating as that was, all I could think of was my future. I was no longer worried about what they had done; I was worried about being charged and put into prison for something I did not do.

I knew that they would lie to protect themselves and back each other up on anything I accused them of and there would be nothing I could do. I knew that nobody in the brig, which was run by Marines, would care or listen to me and to them I would be seen simply as a troublemaker and potential criminal, and because of that would be treated as such. Looking at it from their perspective, my command made the right choice by putting me in the brig. It saved their asses and their jobs because had I been taken to the hospital that morning, I would have sang like a canary.

I was as meek as a lamb when I was brought into the processing room of the brig on Camp Lejeune. They acted like I was a criminal and gave me the full Hanibal Lector treatment. They stripped me and threw a dusty orange jumpsuit at me and then snapped a mugshot. They then chained me up in full restraints as if though I was some rabid serial killer who had just busted on a killing frenzy. I had two sets of cuffs around my ankles and two around my wrists which were also secured to a belt around my waist. They even put a weird chain around my neck which they used to pull me through the complex like a dog.

When one thinks of a jail, they think of rows of cells, a common area, a cafeteria of sorts…well, that is not where they put me. They put me in something reminiscent of what Hanibal Lector in *The Silence of the Lambs* was held in. Nothing but concrete: a concrete slab, concrete floors, concrete toilet, concrete everything; and a bright, cornea-burning florescent light which never went off. There were no windows and no way for me to tell time.

Before they put me in the cell, they stripped me of the orange jumpsuit I was clad in and made me get in butt-ass naked. Once inside the cell, they threw me a pair of used tighty-whities which, to my educated eyes, appeared to have a shit stain in the seat of them. I remember thinking, *it really cant get any worse than this!* I was then left alone in isolation with no idea why I was there, how long I would be there, or what was going to happen next.

I spent the first two days curled up in the corner of the cell trying to shield my eyes from the bright light. It was very cold in the cell and next to impossible to sleep. I never knew what time it was and only slept in short bursts – 30 minutes here and hour there. They would occasionally throw in weird, disgusting food – bread and a red liquid – through the door and I barely ate any of it. The whole time in that cell I kept as calm as I ever had in my life. Looking back, I am shocked at how calm I remained. I never shed a tear or raised my voice once.

My arm was still in pain, but there was no point in asking anyone for help because not a soul in that brig would listen to me. The guards were incredibly apathetic and seemed as though they did not care about anything at all. At one point during the second day the guards shoved in a hose and gave me a "shower". *What the hell is going on around here?*

Is this fucking torture?

Around the third day, I was beginning to unbutton a bit and was even talking to myself, saying things under my breath like, "Please let me out," and "I'm starting to lose it!" I tried to get the attention of a passing guard during some point on the third day, and he actually answered a few questions. I remember asking him if he could please tell me why I was there – like what "charges" they were holding me on because I still had no idea. He hesitated and basically said that he couldn't for whatever reason. It seemed weird to me that nobody could answer a simple question. It was almost as if they had nothing to tell me and I was being held in there for some weird psychological purpose – like a tactic to break me down.

On that same day, I was taken out of the cell, in chains of course, and dragged to an office where I spoke to a "doctor"who looked just like the Kentucky Fried Chicken Colonial. The guy was not trying to hear anything I had to say and only asked me if I was suicidal or homicidal. I think he also asked me if I needed any medications to which I said no to all three. I remember telling him, very calmly, "I don't know why I am in here, Sir. I mean, look at me! I'm chained up like I murdered someone, and I literally have not done anything. I feel like I'm being held against my will." He did not say anything back to me and only looked at me apathetically shaking his head. After the meeting with the "doctor", they took me back to the cell and made me strip again and there I spent the next day curled in the corner trying to sleep.

The following day, I was able to speak to another guard who told me that I would see a judge in a couple days, and they would read for me the charges against me and also set a "bond". Now it was really sinking in! Those fuckers were literally going to try to charge me with something that I did not do. I was beginning to internally freak out about the prospect of being charged and began thinking of legal actions that I potentially might need to take.

All throughout the day, the guards kept making rounds to my cell and took turns throwing things in at me, laughing at me, and taunting me. I kept calm the next two days, but had my mind made up that on the 5th day, I would lose my shit completely. I couldn't take any more of that cell – not one more day.

On the 5th day, right before I was about to let all hell loose and freak out, a guard came to my cell and opened the door. He said, "Verlice. PR!"

I was frozen!

"What does PR mean, Sir?", I questioned.

"Permanent release."

"What does that mean? Does that mean I don't have to see a judge?"

"No," he responded, "your command is here to get you and there are no further actions required." I felt a thousand pounds lift off my shoulders when he told me that any "charge" was being dropped. All I wanted to do at that point was get the HELL out of that torture chamber.

They made me put the jumpsuit back on and for some reason they chained me up one last time and dragged me to the front where I had been processed five days earlier. That time, my interactions with the personnel were a bit different than it had been when I first arrived.

I remember when I got into the processing room, the Sergeant in charge looked at me in shock and screamed at the guards dragging me, "Why the fuck is he chained up like that! Get that shit off him right NOW!" The guards promptly removed everything from me and then disappeared back into the bowels of the brig.

It almost seemed to me that they were privy of something that I was not; it was like they knew I was in there for some phony reason and was being treated unfairly as a result. I signed something and put back on what I was wearing when I was processed in – a green pair of skivvy underwear. I left the brig with a smile, probably the only one to ever do so.

CHAPTER 11

THE RESURECTION

BACK TO THE COMPANY OFFICE

I was driven back to the company office and while I was a bit worried about going inside part of me was also relieved that I was out of that cell; so much so that it didn't even really bother me to face my assailants again. It was around noon when I entered the company office, and it appeared that everyone was gone for lunch so it was just me alone waiting for Satan himself to pay me a visit. I sat at a table for several moments until First Sergeant Kellie (Satin himself) came walking in. He noticed me sitting there and smiled, proclaiming in jest, "Goddamn, Verlice! You look well rested!" I didn't say anything to him but simply smirked back awkwardly.

"You gonna do everything we tell you to do from here on out?"

"Yes, First Sergeant!"

I didn't hesitate in my response and simply wanted to get away from him and leave the office. Their plan had worked; they had finally broken me completely. Those last five days had scared me green, and there was no way I was going to go through that again under any circumstance. If they had demanded to shove something up my ass right then, I would have done it with a smile on my face. Well, okay...maybe not a smile (but you get it).

"Good.", First Sergeant Kellie exclaimed, smiling like a bully on a the school playground. "Now, get out of here!"

"Aye, First Sergeant.", I said, leaving the company office immediately. No part of me wanted to spend any extra time lingering in that office where I was clearly a target, so I went back to my room and put on a clean uniform. My shoulder still throbbed in pain and continued to for the next four months. I slept in a makeshift sling to ease the tension until it no longer hurt.

Many of the guys in the company would still not look at me or speak to me after that incident and I have no idea what type of rumors were floating around, but I'm sure they weren't good. By that point, I did not care in the slightest. My only goal was to get out of the Marines and forget about everything that had happened in the past year.

Whether I wanted to believe it or not, I would be scarred for life by what had happened during my time in the Marine Corps. I tried to pretend that I wasn't bothered by any of it and even walked around in denial; but looking back, it was clear that I was broken. I remember sometimes, some of the Marines would "joke" around and imitate the First Sergeant by screaming my name from a different room. I would run in, all sweaty and panicky, only to see them all laughing at my response. To them it was funny – a joke. To me it was trauma and at that point in time I had no idea how deep that trauma ran.

After that last incident, my discharge process began to speed up and things were getting done in terms of my paperwork. The First Sergeant and even Staff Sergeant Billings' attitude changed towards me from contempt to appraisal. One day, a couple months after the incident, First Sergeant Kellie called me into his office and sat me down. He talked to me like a father and tried to explain things to me. He told me that they were initially going to "court martial" me but because I was such an upstanding Marine, they "worked tirelessly" to have that reversed and let me walk free and clear.

Looking back, he was trying to make it seem like he worked for me so that I would be sure never to tell anyone what happened. I'm quite sure of this because there was nothing they could have court martialed me on. Even if they had tried, and even if "hypothetically" nothing else had occurred in terms of the assaulting which took place and I had simply been a beligerant Marine, the only thing they could have said was that I refused to participate in the battalion exercise. That was not really something one could be court martialed for and would have only warranted a company level non-judicial punishment (NJP) which was basically a just a slap on the wrists. They were scared that I was going to exercise my rights to speak to someone about what took place, but they groomed me and used fear and coercion to keep me quiet.

Unbeknownst to me at the time of the incident, there actually had been witnesses to some of what occurred, especially with the body slam onto the concrete and the knife fiasco. One of the Marines who saw it happened to be a good friend of mine at the time and I can remember him telling me how he saw what happened. I had asked him if he would testify at a court martial should he need to and he told me, "Absolutely, I hate Staff Sergeant Billings." First Sergeant Kellie knew people had seen what happened and if I had decided to say, "Fuck you, First Sergeant. Give me a court martial!", it would have been game over for him and Staff Sergeant Billings and that would have certainly costed them their jobs (and retirements which they were both trying to work for).

To put this into perspective, let me tell you a quick story. A few years after I got out of the Marines, there was an incident with a Sergeant Major who was stationed on Parris Island. He was one of the Depot Sergeant Majors (a pretty big billet to have) and was therefore subject to much scrutiny in terms of how he handled himself (remember, the higher the rank the more of a standard they were held to).

One day, the Sergeant Major was filmed taking a hat from a protester who was standing outside of the depot's gate. All he did was take the guy's hat and yell at him; he did not touch him, hit him, or even threaten to hit him. Just because of that though, they arrested the Sergeant Major and ultimately forced him out of his position and into an early discharge. That is how hard they came down on higher ranks when they fucked up.

So, given what happened to me: the strangulation, the knife, and all the assaulting which occurred over months — had all that went to a court martial, *and* with witness testimony, those two would have been absolutely fried and the USMC would have came down on both of them very hard. They knew that and I knew that, but I was too intimidated to do anything about it. Like I said, they had broken me down to nothing and all I cared about was getting the hell out.

In his office that afternoon, the First Sergeant looked at me and declared, "Verlice, you have balls!" he continued as I sat there without saying a word. "You are one of the only Marines in this company worth a damn, you know that... I mean it!"

He literally said that to me after everything he had put me through, and it appeared to me, at the time, that he meant it! I could not wrap my head around it and still, to this day, I think of him saying that to me and it keeps me up at night.

The only reason he would say that, I believe, is to keep me from telling anyone what happened. He was trying to boost my ego after all they put me through and manipulating me to sympathize with them or to view them as my saviors. Still, to me it was the closest to any apology I ever got from him.

"You just want to get the fuck out of here, don't you, Verlice?" he questioned as I stood rigidly at parade rest.

"Yes, First Sergeant."

That was the last time First Sergeant Kellie or Staff Sergeant Billings ever spoke to me. I think they figured that I had had enough and "bled" enough for my discharge and ultimately left me alone.

After that day, the rest of my time in the USMC was a breeze.

I was not messed with again by anyone for any reason. Perhaps the First Sergeant had told everyone to leave me alone and that I had had enough. Whatever the reason, I appreciated it greatly after everything I went through because I don't think I could have taken anymore. I spent the next few months just existing and spending every weekend with a girl I had met off base. She was probably one of the lowest quality girls that I could have been with, but I was beyond broken by that point and found it hard to talk to anyone let alone a pretty woman.

My last day in the United States Marine Corps was on April 1st. I remember thinking it was funny because I could almost see them saying, "April Fools, mother fucker! Back to the company!" I was so stiff with anxiety, even in those last moments of my USMC career, that I actually beleived that they may pull that move on me that morning.

I woke up really early on my last day and headed over to the company office where I had to get one last signature from the captain. It was the final signature needed on my discharge documents before I could head over to the admin building and get my DD-214 (discharge paper stating one has ended active-duty service). I remember waiting on the second floor of that company office building for what seemed like an hour until finally he came out to sign my sheet. I remember the Captain, who I was unfamiliar with, asking me, "where are you going to go now, Verlice?"

"Home to Florida, Sir."

He nodded, "That's a long drive!"

"Yes, Sir," I responded back.

I remember the last thing he said to me was, "When you go home, don't be telling any of these war stories of yours." He was sort of half joking but also half serious at the same time.

"I won't, Sir".

He paused and then smirked, "Yea you will.", he chuckled before continuing, "Just remember that you were a Marine and you upheld honor, courage, and commitment. The Marines did a lot for you, Son. Just always remember that when you go to tell your stories."

I looked at him and all I could think about was how badly I wanted him to sign that paper so I could fuck off out of Alpha Company. "Aye, Sir", I said trying desperately to hide my impatience.

After he signed the paper, I went outside and saw my platoon forming up for morning formation. I didn't even look at any of them as I walked right by and straight to my car. I got into my car and drove to the admin building where I was given my DD-214 and told, "have a good life" by some POG boot behind the desk. That was it... I was done. I looked at my DD-214 – Honorable discharge. *You made it, Chuck. It's over,* I thought to myself as I left the building.

It's fucking over, old boy!

I remember having a fantasy about taking off my uniform and skipping back to my car and producing my DD-214 if anyone started to yell at me for being "out of uniform". I wanted to do that but then I thought against it because I really didn't feel like having one last ass-chewing before leaving for good. So, with the fantasy out of my head, I got in my car and drove out of the front gates of Camp Lejeune for the last time, flipping off the sentries as I passed by. It was something I had fantasized about doing a thousand times over the past year of my ordeal and when the moment came, I found it to be rather lackluster.

Goodbye, motherfuckers, and thank you all for the wonderful memories. Ill be sure to send you a postcard in the mail!

CHAPTER 12

HOME FOR GOOD...BUT WHAT IS HOME?

AFTERMATH

I was completely elated upon first being discharged from the Marine Corps, but it did not take long for me to realize that I was irreparably damaged. I spent a couple days with my girlfriend in Charlotte and then drove home to Florida where my plan was to find a job and try and be normal again. I was only home for two days when I had a huge fight with my mother over something very petty. After everything that had happened and everything I had been through, the prospect of being yelled at by my mother was ridiculous and so I simply left the house without saying a word.

I ended up going back to North Carolina and staying with my girlfriend and her grandmother for a couple months. I found a crappy job at some e-cigarette company and was fired within two weeks for going off on someone for no good reason. No matter how much I tried to run away from what happened to me, I could not forget the Marines and it haunted me every day and night. I was only 21 and felt as though I had lived a lifetime – a far too difficult one. I kept thinking about how I had just been in the Marines dealing with everything I went through and was now sleeping in a bed with a girl trying to live a normal life. It was surreal to me, and I found adjustment almost impossible. I never talked to anyone about any of it – anything that happened to me in the USMC – and tried my hardest to pretend like it never happened.

My character as a person and as a man was deplorable for a few months after getting out and it is amazing that I was not arrested for doing something to someone during that period of time. I was screaming at people, driving recklessly, and treating everyone around me terribly. It was like I was always looking for a fight for no reason. It was like I needed to avenge myself for what happened and for what I felt I never got justice for. It wasn't until I was out for a few more months that the fog began to lift and I then cracked and crumbled into nothing.

I felt like I had nowhere to go and nobody who cared about me. I travelled aimlessly and even ended up in Alaska for a brief period. There were times when I wasn't working or making money and slept in my car and other times when I slept on someone's couch, be it my parents or girlfriend's parent's. After the "high" from the Marines wore off, I realized that there was nothing left underneath; the person I used to be, someone I considered strong-willed and resilient, was dead and what was left was a weak, feeble, and timid wimp. I could not look people in the eyes or form any sort of friendship, especially with other men. My trust in others was zero and I thought everyone was out to get me and cause me harm. I became incredibly paranoid all the time.

Eventually I ended up working the night shift at a hotel by myself for about a year and that is where I really started to contemplate all that had happened to me in the Marines. I began to see things as they were; however, I remained in denial for many more years. When I got laid off from that job, I decided to go back to college and use the GI Bill. It was during my second semester freshman year that I began struggling with insane panic attacks which only got worse as time went on.

I remember struggling to sit in class one day when I overheard some kid behind me talking about his time in the Air Force. I turned around, eager to talk to another veteran, and told him I had served in the USMC as a rifleman. After several moments of talking, I had to excuse myself from the class because I was feeling a panic attack come on. On my way out, I heard the guy whisper under his breath to the guy sitting next to him, "damn, what did they do to him?" It was obvious to me that everyone could see how outwardly damaged I really was.

Around that time, I tried getting help from the VA but was unsuccessful and even went to a VA counselor to whom I told some of what had happened. He said it was awful but again could not really help me as he could not prescribe me any meds. I really needed something to calm me down because I was literally to the point of shoving my head into muddy water behind gas stations. I left school after a year and struggled with homelessness for a while. I also continued to sleep on my parents' couch for a while here and there when they would allow me to. I went bankrupt, lost my relationship, and only crumbled further as I continued to bottle everything that happened up inside of me.

Around 2017, I figured that I didn't need the military or VA and that I would just do it all on my own. I was set on dealing with my problems alone with no one's assistance and that is when I picked myself up, even through the anguish, and continued with life. I went back to school and got my bachelor's degree from Florida State University and then my MBA from the University of North Florida in 2022. Since then, I found work at an international bank that allows me to work from my home which I purchased in 2022. I don't think I could work from an office anymore – at least not now as my panic attacks, anxiety, and depression are so bad that I even have a hard time going to the gas station on most days.

Things have changed in my life and although it appears, to the ignorant eye, that I have overcome all my issues, I have not...not at all. Not a day has gone by since I have been out that I don't think of everything that happened while I was in the Marine Corps. I go over it again and again in my mind like a crazy person. It just won't leave me; it just doesn't go away. It was only as of recently that I came to accept the fact of what Master Sergeant Sneed had attempted to do to me. I was in so much denial about that for years that for a while, I had convinced myself that I had deserved it.

It's almost as though I am still there, just waiting to go to formation or hear First Sergeant scream out my name. I see that Marine vomiting blood and understand now why I can't eat enough to ever gain weight. I think all of it will follow me around for many years, and maybe even the rest of my life. It is part of who I am now – part of me.

CHAPTER 13

WHAT I THINK *REALLY* HAPPENED

Why the things that happened to me in the Marine Corps ended up happening, I'll never truly know. I'm not sure if Master Sergeant Sneed was a closeted homosexual, frustrated at the Military's "don't ask don't tell" policy, or if he just wanted to go on a power trip that morning. Either way, I really don't think that I deserved any of the things that happened since I had done absolutely nothing to deserve it. I really feel as if though I was cheated out of my USMC experience and that had Master Sergeant not attempted sodomy on me that morning, I would have continued on with my job as normal.

It was directly because of Master Sergeant Sneed that First Sergeant Kellie got involved and then instigated all the abuse which followed. I'm not sure if First Sergeant Kellie even knew why I had been in the office with Master Sergeant Sneed to begin with or if Master Sergeant Sneed had even told him anything of what *really* happened. I seriously doubt it. To the First Sergeant, I was just insubordinate or belligerent and deserving of whatever he wanted to dish out to me. Either way, he did not conduct himself in the way a high ranking Marine was "supposed" to. Even if nothing had happened that morning when Master Sergeant Sneed approached me and I had simply just been struggling with something internally, he should have shown me some type of compassion towards another Marine. He should have conducted himself like Sergeant George had back when I was in infantry school. That is what *I thought* Marines were supposed to do. The truth is that I wasn't asking for compassion, leniency, or grace; I just wanted to be left alone.

I think, looking back, that First Sergeant Kellie wanted to ensure that I didn't just get out of the Marines so easily. He wanted to make sure that I would not leave without a scar, be it mental or physical. To him I had earned nothing and would have always been a "boot" in his eyes no matter what I had done. First Sergeant Kellie was a decorated Marine and likely thought that just because I had not seen or done the same things he had I was therefore less than he was as a Marine. It seemed to me that in his mind he thought that there was no possible way I could have any issues. That was the mentality in the USMC infantry back then. No mercy given for any situation.

How much of what happened was a plan? That I will never know. I'm sure they planned it to some extent; it seems too perfect. When all arrows point to something, it usually indicates that it's correct, but since I haven't spoken to any of them, I'll never really know for sure. Who was all complicit?

Who knows?

Did the CO of the company know and approve of certain things that happened? I think he probably did, at least to some extent. He at least made sure no officers were around that morning when both First Sergeant Kellie and Staff Sergeant Billings assaulted me and then most likely helped them plan to keep me quiet afterward. He obviously had their backs and wanted to make sure that *they* didn't go down for what they had done to me.

Little things happened after the last incident which indicated to me that bigger plan had been put in place. For example, I remember one evening before Christmas, the CO called me into his office and essentially warned me, in a polite manner, not to go to the hospital for any reason. I remember wondering why he was telling me not to go to the hospital. It was as though they were afraid that I was going to talk to someone there. I know that; I feel it in my soul. I remember he asked me what I was planning to do if I was feeling suicidal and my response was, "...go to the naval hospital, Sir."

I told him what I thought he wanted to hear because that seemed like the most normal answer. I remember he quickly told me, "...NO!..." and that if I was to feel suicidal that I was to immediately call my Sergeant. He made it out like it was because he "cared" – because I was his Marine and he wanted to make sure I was taken care of. The real reason, I believe, is because he *knew* if I went to that hospital that I would end up telling someone over there what happened, and an investigation would have been placed on the higher ranking members of the unit.

Another giveaway to me was the Sergeant Major. After I had been brought before the Sergeant Major of the unit the day of the attempted drowning and he had instructed nobody to touch me again, First Sergeant Kellie would had to have been very careful in making sure that the Sergeant Major did not know of anything going on moving forward. If the Sergeant Major found out that they had continued the abuse and assaulting after his direct order was given not to touch me, it would have been a death sentence for all those involved. The Sergeant Major of my unit did not play around and everyone in the unit knew it.

When they put me in the brig for whatever they claimed I did that morning, the Sergeant Major and BC both would have had to have been contacted and made aware that one of their Marines was being confined. Being put in the brig is a major incident and something that would not have slipped past the Sergeant Major and certainly not the BC of the unit. Furthermore, if they were really going to court martial me like First Sergeant Kellie had said that day in his office, the Sergeant Major and BC would both have known about it and I would have certainly gone before both of them for the incident. That never happened; I never went before either of them.

I saw the Sergeant Major in the hallway of the company office a couple months before I was discharged (after the whole brig incident had occurred) and he remembered me from the only time I had been in his office. He looked at me and said, "Verlice, you look so much better than the last time I saw you. Glad I haven't heard anything more about you since." It was very clear to me right then that he never knew of anything that happened; he knew nothing about the incident or me being put into the brig. That let me know that everything the First Sergeant told me about the "court martial" was likely bullshit and spun up to cover their own ass and keep me in line. I knew that but still refused to report any of it to anyone. In all reality, all I had to do was walk into the Naval Hospital on a Saturday and report it all and that would have been it.

The entire situation was against the rules to begin with as hazing in the USMC, even at that time, was zero-tolerance and if a Marine was to step out of line, the higher ups – Sergeants, Staff Sergeants, First Sergeants, and COs – were supposed to go about the proper avenues of disciplining said Marine, usually through administrative proceedings (NJP, court martial, etc.). Even if no plans had been drawn up and I had simply been a bad Marine (which I was not), they would have still been in the wrong for assaulting me the way they did for as many months as they did. That was not the way anything should have ever been handled, even in the Marines. Even though hazing happens amongst the lower ranks, it was exceedingly rare for it to occur with higher ranking Marines. The whole situation was unique in the worst possible way.

The irony is almost comedic given the fact that the First Sergeants and Staff Sergeants, the same ranks as those who instigated and perpetuated all the assaults on me, were the very ones giving the weekend safety briefs to all the Marines. The safety briefs consisted of warning everyone to avoid trouble in town, not to drink and drive, and of course…not to haze or abuse other

Marines. First Sergeant Kellie gave those speeches so many times and yet assaulted me regularly during the time I was in the Marines. It was baffling to me, and quite frankly it still is. How can you ever trust again? Who was I supposed to go to when the highest ranks in my company were the ones doing all the assaulting? These are questions I ask myself everyday – still.

I can look back on the situation now with a clearer head and see what happened, but at the time I was unable to logically make reasonable decisions. I was physically able to handle the Marine Corps, that wasn't the issue for me; I could even handle the mental games and the mind-fucking they sometimes gratuitously threw at us. None of that was the problem for me.

What it boils down to –it's base element – is that I was not socially mature enough to handle the type of abuse which I was subjected to. I grew up very isolated and had very little experience dealing with people, even in a normal capacity; now being in the Marines adds in the stressful living environment and mentality of Marine Corps life where things were not approached or handled in a normal way akin to what one would expect in any other civilian job. With that in mind, one can see why it was not possible for me to handle the situation as "others" may have in a normal environment.

They had manipulated me in the most intricate way; they had manipulated me, in my opinion, out of fear for their own jobs and freedom. Everyone says the Marine Corps is full of dummies, but I disagree. They were incredibly clever with the way they went about handling me. They abused me to no end, isolated me in the brig, and then immediately upon letting me out, they left me alone.

They let me cool down for a couple months and then brought me into the office, said wonderful things about me, and talked about a court martial that was, in my opinion, never going to occur. They had, rather surreptitiously, shifted my focus from the assaulting, abuse, and illegal confinement to my own freedom.

I had walked away from that last meeting with the First Sergeant feeling relieved, as if I had been spared the guillotine for something "illegal" I had done. I even went back to one of my friends and boasted about the "good deed" First Sergeant Kellie had done for me and the words he commended me on. I had actually felt, in that moment, as if I had deserved a court martial for what they had done to me. I appreciated them and venerated them for their leadership. Some, I have been told, call this Stokholm Syndrome and I had never realized any of this until just now as I'm writing this.

When everything went down the way it did, I simply lost myself at that moment in time. I didn't know how to handle it or how to respond to it. I could have reported it and perhaps, given the circumstances, I would have been given a different duty assignment or change of unit had I persisted, but I simply did not have the logic I do now to make those rational decisions. It's easy to look back and say, *I should have done this*, but at the time, again, those options were not available to me because I was too brainwashed to tell anyone. I had simply shut off and went into survival mode where nothing else mattered except getting away from the abuse.

I have not seen or talked to any of them since I have been out, but sometimes I wish I could. Sometimes I wish that I could talk to the First Sergeant, if he would even remember me, and ask him why he did it. I would ask him if it was a plan and what the purpose of it all was. I would tell him of the toll it took on me and the price I paid and continue to pay to this day. Knowing the person he was, I'm sure he wouldn't care and would probably laugh in my face. I don't know if either of those three involved remember me, but they should know that I remember them and will *never* forget.

CHAPTER 14

DEEPER THAN THE MARINE CORPS

Everything up to this point in the story has been about what the Marine Corps did to me, and I could easily say that the Marine Corps has been the bane of my existence and caused all my problems…but that would not be one hundred percent truthful. Looking back at my childhood, I can see now a lot of what I couldn't see then. It was my childhood that ultimately led me to the Marine Corps and led me to my Waterloo.

My childhood was not profoundly bad by any means, but it also was not super stable either. Neither one of my parents was super attentive or doting, and my father in particular was quite apathetic. I don't say this to bash on my parents – none of it was their fault – but it is instrumental in that it caused me to form certain ideals which led me to my fate.

As I grew from a young boy to a teenager, I began to develop negative thoughts which could have easily landed me in an anti-depressant-laced nightmare, but because I was so used to dealing with things on my own and was not helped in anyway be either one of my parents, I build a façade of resilience and strength which I believed was real.

Of course hindsight is 20/20 and looking back, I realize I was not cognizant of the fact that my façade was just that. I spend my time growing up as a teenager completely alone, living in the woods, hunting and learning various survival tactics which only served to build this façade stronger. I ran eight miles a day home from school on the side of a busy highway – rain, sun, heat – it didn't matter. I had garnered a sort of reputation at school for being a "bad ass" and this began to give me a feeling of immortality… like I was untouchable.

I also read a lot and romanticized certain book characters which I wanted to emulate. I began to feel as though I didn't need anything – money or relationships; I thought that I was super strong and self sufficient. I felt powerful and resilient but it was all just a phony wall built of very flimsy twigs. I think, however, had I not espoused this mindset, I probably would not have made it to adulthood. When I graduated high school and realized that I had no plan, it was this mindset that led me to the Marine Corps infantry. My feelings of immortality and strength made me think that a damn bullet couldn't stop me. That was a fatal mistake on my part as I was not socially adept or assimilated enough to take on even the most basic form of society let alone the vicious and hostile environment such as the Marines.

By the time I was in the Marines, I realized that my façade was easily breeched and all the problems from childhood, problems I didn't even know where there or that were buried very deep, began to be dug up and put in the spotlight. The Marine Corps was just the straw that broke the camel's back and by the time they got a hold of me, and everything discussed in this book ensued, I was left with nothing – no walls and no façade. The resiliency and strength, or whatever kept me kicking throughout high school, was dead and gone, and I have never been able to find it again. I guess I was never really that strong or resilient to begin with.

Looking back, and not to get creepy on you, but I believe I was given the option to join the Marines and go through what I did for some greater reason unknown to me. I am not the type to get religious on anyone, but there were several things which occurred prior to me joining that, now to me, appear to be either God or the universe pulling me in a certain direction. Maybe I was given the choice, and I chose the wrong path. Perhaps I failed my test and now must live with the great consequences of that failure.

When I first made my decision to join the Marines, as stated, it was on a whim, and I really did not think the action through with any great depth at all. I remember telling my mother, who was on the side of our small double-wide trailer partaking in some of the Devil's Lettuce (artificial Devil's Lettuce I should state) and was probably not coherent enough to even appreciate the decision I had made. I remember stating to her that I was going to join the Marines and she laughed, proclaiming, "Yea, right!".

It sort of lit a fire under me when she responded in that manner and set the ball in motion. I also told my father, who again had served for over 20 years in the USAF and was in fact the only one who tried to talk some logical sense into me. I did not listen.

About a day later, I had come to my own senses and decided to mull it over a bit more; however, my mother had remembered what I had told her and took the liberty of having a USMC recruiter call me, randomly, during the middle of the day. Because my mother was so adamant that I left the house as quickly as possible, I felt obliged to go along with my hasty decision to join the Marine Corps. A day later I was at a local library talking to the recruiter, listening to his car salesman-type speech, and a couple days later I was at MEPs swearing in. It all happened so quickly.

After I swore in at MEPS, however, I started to become very gung-ho about my decision, and it would have taken an act of God to stop me from joining. Looking back, I believe God did try to stop me and of course I did not listen. When the recruiter asked me what I wanted to do in the Marines, I told him without hesitation, "Infantry". My mother, who was next to me when I stated this, told the recruiter she did not want me to do infantry, and the recruiter listened to her. He told me that I scored high on the ASVAB and did not need to do infantry. Still, I persisted. He eventually told me that there were no infantry slots available and that I would have to choose a different job.

Over the next couple of months, he told me various jobs that I could do and for some odd reason, none of them worked out. Either he would not respond to me about them when I would inquire, the job would suddenly become "unavailable", or he would simply not produce the contract for me to sign. Eventually, I ended up deciding to choose the MOS of military police, which

I really didn't want to do, but felt I had no other choice because I had to leave quickly. I was resolute on doing that job when one night my mother had gotten loaded on "artificial weed" and booze and decided to text my recruiter in a somewhat flirtatious manner. The recruiter responded angrily and then stated to me that he would continue to work with me on the promise that my mother would not attend any further recruitment meetings along with me. I agreed.

The next time I met the recruiter, as he was weighing me in, he looked up at me and said,

"Verlice…do you still want to do infantry?",

"yes, sir"

"Good, because a spot just opened up!"

Seriously, you cannot make this shit up! A week later I signed the infantry contract which sealed my fate and set me on an irreversable course to Hell. It was like the universe was pushing me towards that path and I was destined to have that experience. I was meant to learn something from all this, that much is clear.

Maybe I'm reading too much into it, but there were four separate dates which I was supposed to ship out that kept getting canceled for some unknown reason and the last time it happened I became impatient and told the recruiter I was going to speak to the Army or Air Force – in a vain attempt to get him to hurry along. I remember him saying that because I had stated that, I may not be "Marine material", and that patience is a virtue. But he also told me that he could void out my paperwork, no hard feelings, and go to another branch. I decided to be patient and a week later I shipped out to Parris Island. Whether that was God, fate, or simply just coincidence, it is something I think about all the time along with my entire Marine Corps experience and find myself asking, *what if?*

CHAPTER 14

WHAT I LEARNED

The moral of the story, right? There must be one! What did I learn from all of this? In many ways the Marine Corps was good for me. I think that had I not joined; I wouldn't be where I am today. It gave me drive and motivation to succeed even in the face of extreme adversity. I learned a lot about who I am as a person, my limits, and what I am capable of when shit hits the fan. The Marine Corps turned me into a man, and I wouldn't trade that for anything in the world. I know myself through and through because of the Marines.

On the other hand, I experienced a lot of hell during my relatively short stint in the USMC. I saw some of the weirdest and nastiest things I couldn't even have dreamt about in civilian life. I saw the worst side of human beings and what people are truly capable of in an overly stressful environment. Most importantly, and perhaps to my detriment, I learned to never trust a soul. Making friends is next to impossible as I find it too hard to relate to anyone. I realized that no matter what happens, I am always alone in this world and only have myself to count on.

The whole experience now seems like a fever dream to me and sometimes I find myself wondering if any of it ever really happened. Of course it did, but I have no body – no friends, former team members, or anyone else – to corroborate or talk things over with. It feels like waking up from some unreal and macabre dream that I remember in its entirety and with impeccably clear accuracy.

Sometimes I feel as if though I had died sometime back in 2012 and this perpetual panic, depression, and self-loathing is my personal Hell. Of course, that is just nightmare talk and the reality of the situation is that years of dissociation, incredulousness, and anger have corrupted my normal thinking channels. I guess in some ways that is the price I must pay for not heeding the warnings which were, I believe, clearly laid out before me. It was my ultimate life lesson!

Over the years I've had to learn to accept what happened and acknowledge it for my own sake. I spent many years blaming myself for what happened; I ran away from it never telling a single soul. I felt like a coward – like a bitch and a traitor. I felt like I let my country down when I decided to leave the Marine Corps.

Looking back, I was none of those things. I handled it the only way I knew how at the time and through many days and sleepless nights led me to the point where I no longer blame myself for the things that happened. I did all I could do with the cards I was dealt and tried my best to uphold my commitment to the Marine Corps. It just was what it was, and no amount of self-loathing can change what happened. I was still a Marine and I am still proud. That is something that can never be taken away from me. I still love the Marine Corps; I still get the chills when I hear the Marine Corps Hym. I still feel the desire to put my cammies back on long after I took them off for the last time. I AM still a Marine and I will be one until the day that I die.

The Marine Corps told me a story: a story of self-sacrifice and self-preservation. I anted up and put everything on the table when I signed that infantry contract just as First Sergeant Kellie, Master Sergeant Sneed, Staff Sergeant Billings, and everyone else I served with had done. I joined knowing full and well what could befall me as a United States Infantry Marine and still I never feared it. My decision to leave the USMC had nothing to do with the contract I signed or the job I was there to perform. It was solely fueled by the unjust doings of a select few - a select few who cheated me out of what I felt was my destiny at the time.

What ended up happening to me during my stint in the USMC was something that I had not accounted for, or thought would have ever happened and had I known it would, I might have made different choices about the job I was going to do. The Marine Corps, however, gave me a start to my life which was very bleak and destitute beforehand. I had nowhere to go and nothing to do. My parents were poor and needed me out, that's why I was there. Still, I find it hard to believe that I went into the Marines with absolutely nothing and came out with so much less. The resiliency and strength I once had dissolved into a mass filled with anger, regret, loathing, and great feelings of injustice.

This story was not meant to be a bash on the Marine Corps. I believe in what the Marine Corps does; they have to train their people hard for the rigors of combat, I get it. But the things that happened to me there had nothing to do with any of that or the job I had signed up to do. Some people join the Marine Corps and love it, ending up with a 25-year career. Some, like me, have a different story. I'm sure the hard chargers out there and the motivated devil dogs who may read this will think I am a coward. All I can say to that is - I signed the same contract you did, devil, and was willing to put just as much on the line as you. I just wasn't going to put up with that level of threats and abuse and I dont think that anyone should have to.

I am still proud of myself for deciding to earn the title of Marine and had I been put in a different unit or chosen a different MOS, perhaps I would be writing a story about all the fun I had during my service. But that is not the way it went. I believe everything happened for a reason and maybe someone reading this will be able to draw some sort of insight or inspiration from my experience. If you don't take anything else from my story, at least take this: think things through when making huge life decisions; have a plan; complacency kills! Always speak up when bad things are happening to you whether at home, work, or school. I wish I had; God, I wish it every day. Perhaps I may have been able to save myself instead of expecting someone else to do it for me.

Young men...NOBODY is coming to save you! Only you and you alone can do that.

Even though I am out of the Marine Corps and many years have gone by, sometimes I find myself back on Parris Island standing at the position of attention around the Iwo Jima monument that morning I became a United States Marine. I can still see myself with my right hand raised as I scream the last line of the oath to my country... *"SO HELP ME, GOD!"*

THE END

AUTHOR BIO

Charles "Chuck" Verlice began writing stories in 9th grade as a way to escape the bordom of school. An avid outdoorsman and distance runner in high school, Verlice was lured to the Marine Corps at the age of eighteen by the adventure, thrill, challenge, and prospect of going to war. After the Marine Corps, Chuck spent time adjusting to civilian life and attended college where he earned a Bachelor of Arts in English and a Master of Business Administration (MBA). Chuck currently resides in Florida with his fiance and dog, Beauty.